Spinning Flight

Spinning Flight

Dynamics of Frisbees, Boomerangs, Samaras, and Skipping Stones

RALPH D. LORENZ

Springer

Ralph D. Lorenz
Lunar and Planetary Laboratory
University of Arizona
1629 E. University Blvd.
Tucson, AZ 85721-0092
USA
rlorenz@lpl.arizona.edu

Library of Congress Control Number: 2005937515

ISBN-10: 0-387-30779-6
ISBN-13: 978-0387-30779-4

Printed on acid-free paper.

9 8 7 6 5 4 3 2 1

springer.com

Preface

This book is about things that spin in the air or in space.

Specifically, it is about things that spin in the air and space that I find interesting. I am by training an aerospace engineer, but work as a planetary scientist. Indeed, as an aerospace engineering undergraduate, I regarded—unfairly in retrospect—my lectures in fluid mechanics and aerodynamics as only a necessary evil on the noble road to exploring space where such subjects tend not to apply. My main project during my 15-year career as engineer and scientist has been the *Huygens* probe. In early 2005 this probe descended through the atmosphere of Saturn's moon Titan, where as it turns out, these fields applied after all. In an attempt to gain familiarity with the dynamics of a slowly spinning vehicle like *Huygens* under its parachute, I began in 2002/03 some experiments with instrumented small-scale models. These little models recorded the swing and spin with small sensors, and provided me with insights I would not otherwise have gained, and not a little entertainment besides.

Some months after these experiments began, and as my instrumentation became more compact, I had the idea (while sitting on an airplane, appropriately enough) that the instrumentation was compact enough to install on a Frisbee without terribly altering its flight characteristics. I duly made such experiments, which introduced new challenges in attitude determination and range instrumentation. I found that there was in fact relatively little published work on the subject of Frisbee aerodynamics. I therefore had the opportunity to make some genuinely new observations, which have since been published in the academic literature.

I also observed that almost everyone I spoke to (mostly scientists and engineers, it must be conceded . . .) thought that these experiments were rather cool. It might be interesting to assemble my experiments with the modest body of scientific work on the subject, although the Frisbee research would not be enough subject matter for a worthwhile book by itself.

However, these investigations reactivated latent interests of mine in many other areas. Like millions of other people, I have marveled at how a boomerang flies, or how a stone skips across the surface of a pond. And I realized that there was a common theme to these subjects—that of spinning flight—and the idea emerged of compiling a book with that theme. Thus motivated, I also began other experiments which are reported here.

Exploiting to the full one of the few privileges afforded to an author, I have been liberal in my interpretation of the theme. Although it was not in the project as originally conceived, I have interpreted "flight" to include space, thereby encompassing certain dynamical aspects of space probes, satellites, asteroids, and planets. I make no apologies for this—these cases are just as interesting as the more classically "aeronautical," and many are more so. Similarly, my coverage of spinning disk-wings such as Frisbees stretched a little to embrace radar early-warning aircraft with spinning disk antennae, and thence to include a few words and pictures on nonspinning disk-wings.

Where I have drawn a line—one must be drawn somewhere—is before rotorcraft. Helicopters have fascinating aerodynamics and gyrodynamics, and are magnificent machines. There are also many excellent textbooks and more popular works that cover them in detail—I certainly have no significant insights of my own to add. The one exception is a class of rotorcraft wherein the whole body of the vehicle is spun up, in addition to the rotor. Again, the criterion for inclusion has been that I thought this was novel and interesting.

The book is not intended as a textbook, although students and researchers in various fields may find ideas for many outstanding investigations or problems, and I have tried to be rigorous in my use of terminology. Equations (simple ones) have been used in the text where they are the most succinct way of expressing something, but I have no wish to deter the casual reader.

In the hope that readers may be motivated to pursue investigations of their own, I have included appendices with some technical details of my own experiments and have been fairly rigorous in including references to papers. None of the bibliographies can claim to be complete, but should give ample starting material, and certainly are representative of what I have found to be the most recent, comprehensive, or useful papers on the various subjects. Although it did not exist 15 years ago, it barely needs stating now that the Internet is an enormous resource for information. A web search will rapidly bring far more material than is in these pages.

The book before you is a little larger than the original outline proposed to the publisher, and I thank them for their indulgence in accommodating the additional material. This is the first book I have written myself (I have had the good fortune to write a couple of previous books with some excellent and experienced co-authors) and so in this case all errors are entirely my own responsibility.

As this book was being completed, the *Huygens* probe ended its long journey, successfully parachuting down to the surface of Titan. As I and my colleagues try and understand the probe's behavior and its environment, I have found the intellectual preparation deriving from my experiments to have been quite useful. To play is to learn.

Acknowledgments

Harry Blom at Springer has been an enthusiastic supporter of this project—his strong commitment from an early stage has been instrumental in getting this book done. Some other editors, Ingrid Gnerlich at PUP and Peter Gordon at CUP, have been kind enough to provide early feedback on the idea.

Zibi Turtle, Joe Spitale, and Jess Dooley have all helped with field experiments. Helen Fan and Melissa Myers have provided much-needed help with digitizing data and drawing figures. Some of this activity, together with digging reference material out of the excellent University of Arizona library, was supported by NASA through the Arizona Space Grant Consortium. Melissa Myers also helped greatly with some proofreading and editing.

I would never have started on all this without inspiration from the comprehensive websites on Frisbee dynamics by Jon Potts and Sarah Hummel. Similarly, the excellent Wx-Bumms software made available on the web made boomerang simulation come alive easily.

I thank Erwan Reffet for drawing my attention to his work on balls on cylinders, and Kelly McComb for information on the Turboplan. Raytheon Inc. provided information on the Whirl, and Textron Systems Inc. was kind enough to review the text on Samaras.

Special thanks are due to Alan Adler for taking time to explain the history of and providing material on the Aerobie and other toys. Ted Bailey was generous with his time in explaining some boomerang background and photographing some of his boom collection. Mike Bird explained much of the background to the Giotto encounter.

Obtaining permission to use graphic material is always something of a chore for an author. Credits for images are given in the relevant captions, but let me record here my appreciation for the prompt and generous cooperation from the relevant parties.

Brian Riddle at the library of the Royal Aeronautical Society, and Claudia Condry of the Imperial War Museum, both in London, are thanked for their assistance in the archives.

Lastly I thank the friends and colleagues, and not least my wife Zibi Turtle, for patiently tolerating my ranting about spinning things, and indulging my experiments and writing.

Contents

List of Figures

CHAPTER 1

CHAPTER 2

CHAPTER 3

CHAPTER 5

CHAPTER 6

CHAPTER 7

CHAPTER 8

CHAPTER 9

CHAPTER 12

CHAPTER 13

APPENDIX 1

APPENDIX 2

1 Introduction

Why does a Frisbee veer to the left at the end of its flight? Why does a skipping stone veer to the right? Why is a golf ball dimpled, and how can it jump out of the hole without bouncing on the bottom? How can a discus be thrown further upwind than downwind? How does a boomerang work? How can the spin of an asteroid let us measure its strength, and how can the spin determine whether it will hit the Earth? All these things are connected.

This book is about spinning, flying things. Because I found many related topics to be interesting, it also discusses a lot of other spinning things and flying things, and while my original intent was to consider flight only in the sense of aerodynamic flight, I was drawn to include many aspects of spin in spaceflight too. It is the similarities and contrasts between these different situations that appealed to me.

It is assumed in this book that the reader is familiar with some basic aerospace terminology and principles of vector mechanics. However, it is useful to review certain definitions and ideas, especially since the scope of this book is rather broad, covering some astrophysical topics as well as gyrodynamics and aerodynamics. Advanced scholars may justifiably feel brave enough to skip ahead—if matters become confusing, they may always retreat to this introduction.

For rigorous explanations of dynamics, the reader should consult a mechanics textbook, of which there are many excellent examples. There are texts large and small in which even the deepest thirst for vector algebra and calculus should be satisfied. Although this book is basically all about principles and ideas, I've allowed a few algebraic expressions into the text because I think quantitative expressions are a good thing, and often the most succinct way of writing the relationships between quantities is an equation. But you don't need to worry about them—it's the principles that count. If things sort of make sense in your head after reading this, then it has achieved what I intended. It is my hope that some readers may even be inspired to study some of these spinning phenomena on their own. I'd love to hear the results.

Basic Gyrodynamics

The spin of objects has analogies with the motion of bodies through space—Newton's laws apply. Just as the (translational) momentum of an object will remain constant with time unless an external force is applied, so the angular momentum remains constant unless there is an external torque. Spinning bodies have a property analogous to mass, called "moment of inertia." The moment of inertia about some axis is equal to the integral of the mass times the square of the distance from that axis: for a pair of masses m held apart by a light rod of length 2∂, the moment of inertia about the center is $2m\partial^2$. For a thin disk of mass m and diameter ∂, the moment of inertia about an axis through the center is $m\partial^2/8$, while about a diameter in the plane of the disk (i.e., a transverse moment of inertia), the moment of inertia is $m\partial^2/16$. An object with larger moment of inertia is harder to get spinning at a certain speed—it requires more energy.

The surprising dynamic behavior of spinning objects is easy to understand with vector mechanics. A spinning object has an angular momentum vector which points along a particular direction in space (a convenient mnemonic is the "right-hand rule"—if the curled fingers of the right hand point along the sense of rotation, the thumb indicates the direction of the angular momentum vector).

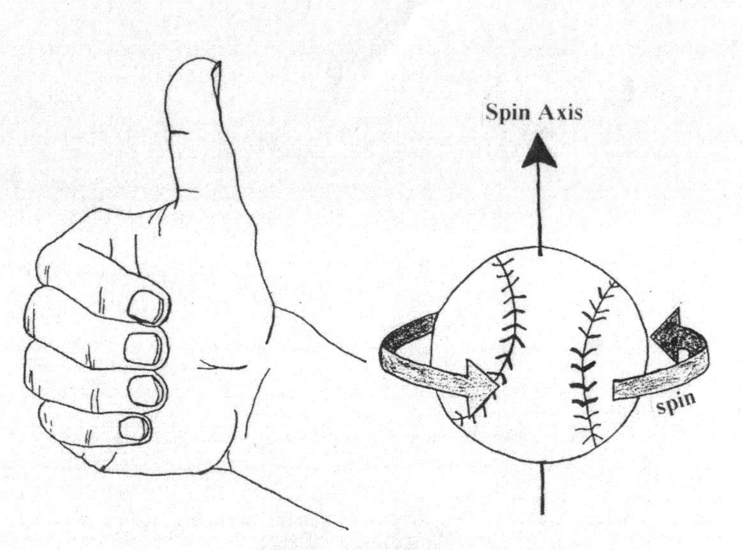

Figure I.1. The right-hand rule—the thumb defines the spin vector where the fingers show the sense of rotation.

Unless an external torque is applied, the vector remains constant in magnitude and constant in direction. This is the principle of the gyro compass. But if an external torque is applied for a short time, the torque–time product gives an incremental vector that is added to the body's original angular momentum vector. The two angular momentum vectors are added, as if the "tail" of one arrow were placed at the point of the first: the resultant vector is simply the line joining the tail of the first to the point of the second.

Let us consider a classical toy gyroscope—a wheel mounted in a frame. If this wheel is set spinning in a horizontal plane such that its angular momentum vector is vertical, it is symmetric and there is no external torque due to gravity, and the wheel stays fixed in a vertical orientation.

But now spin the gyroscope up in a vertical plane, giving a horizontal angular momentum vector say, pointing right and support the frame its right end. Now, the weight of the gyroscope acts at its center of mass, while the reaction force balancing that weight is acting at the support. There is thus a torque being applied (in this case, the torque is anticlockwise, and thus the torque vector points out of the page). In a small interval of time, this torque adds to the angular momentum vector—rotating it from due left to slightly out of the page. Thus a gyroscope mounted this way precesses anticlockwise as seen from above. (The precession direction would be opposite were the wheel spinning in the other direction.) This nonintuitive behavior, namely the precession of the spin axis in a direction apparently orthogonal to the applied torque, is the key feature of gyroscopic motion and is essential in the understanding of the flying objects in this book.

The angular momentum of a rotating body is equal to the product of its moment of inertia and its angular velocity ($I\omega$). On the other hand, the rotational kinetic energy is equal to half of the product of the moment of inertia and the *square* of the angular velocity ($0.5I\omega^2$). Unless external moments are applied, the angular momentum of a system must remain constant. The canonical example is of a figure skater who draws in her arms while in a spin: drawing in the arms reduces the moment of inertia, and so for angular momentum to be conserved, the spin rate must increase.

In general the mechanics of rotation can be described by a set of expressions known as the Euler equations. The inertia properties can be represented by a tensor (a matrix of 9 numbers), but for most applications only three numbers (and often only two) are needed: only the three diagonal terms in the tensor are nonzero. These are the moments of inertia about three orthogonal axes, the so-called principal axes. The principal axes are the axis of maximum moment of inertia, the axis of minimum moment of inertia, and the axis orthogonal to the other two. It can be shown that stable rotation only occurs about the minimum or maximum axis—a result sometimes called the "tennis racket theorem." You can easily verify this—it is easy to spin a racket (of any sport; it applies to cricket bats too . . .) about the long axis along the handle, or to toss it like an axe, so that it rotates in the plane of the strings. It is impossible for it to make sustained flips end over end in the orthogonal direction (which is the "middle" moment of inertia).

Although it is possible to spin an object about its long axis, this motion is not necessarily stable in the long term, a result that was learned the hard way in the U.S. space program (see chapter 5). Specifically, internal energy losses force the object to ultimately rotate about the axis of maximum moment of inertia.

However, internal energy dissipation (for example, flexing of imperfectly elastic elements or flow of viscous fluids) can absorb energy. If an object is spinning about an axis other than that of maximum moment of inertia, and energy is dissipated, the system must compensate for a drop in angular velocity by increasing the moment of inertia. In other words, it rotates such that the (constant) angular momentum vector becomes aligned with the maximum moment of inertia—the only way of reducing energy while keeping the angular momentum constant is to increase I as ω decreases. The rotation about the axis of maximum moment of inertia, the stable end state, is sometimes called a "flat spin," since the object sweeps out a flat plane as it rotates.

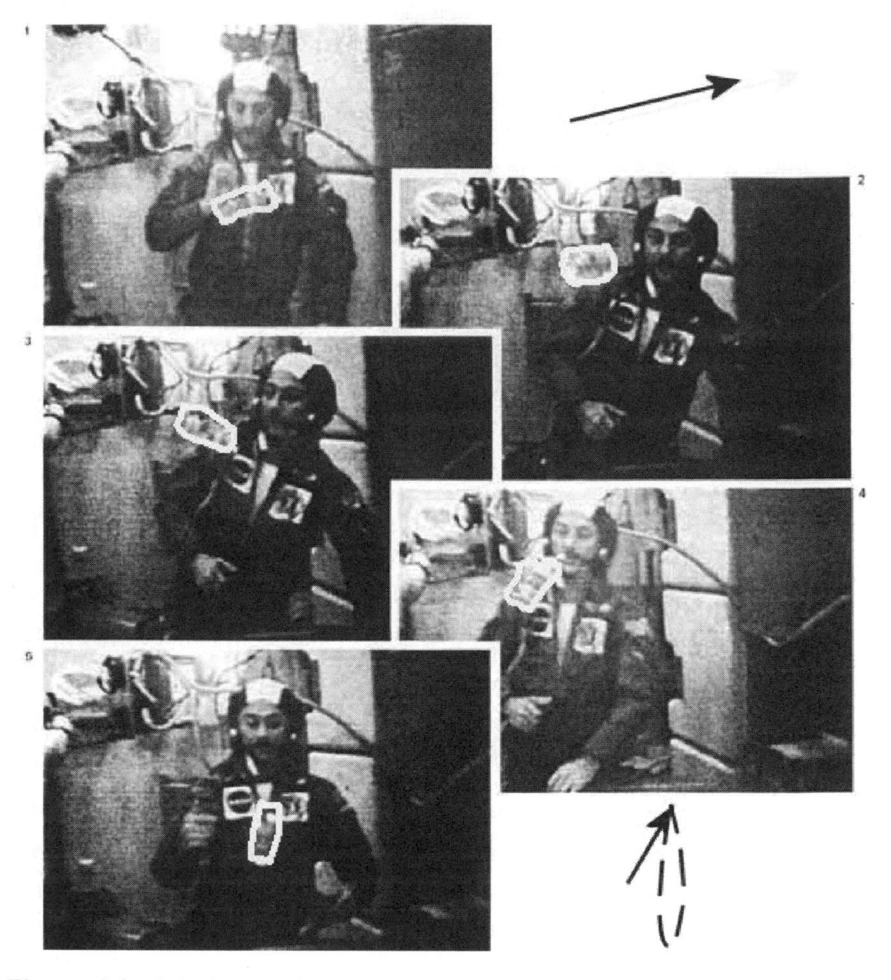

Figure 1.2. Spinning the Bottle. Astronaut Owen Garriott on board the space station *Skylab* in 1972. He set a bottle of water (highlighted in white) spinning around its long axis, as indicated by the upper left arrow. Within a couple of minutes the dissipation by the liquid had caused the bottle to begin describing a cone (lower right arrow) and finally to turn end over end in a flat spin. The angular momentum vector remains the same in magnitude and direction at all times. NASA images.

Two terms are often used (and misused) in connection with rotational dynamics. The first is *precession*.[1] This applies to the movement of an angular momentum vector by the application of an external moment.

---

[1]Note that in geophysics—as in the precession of the equinoxes—this means something different.

The second effect is *nutation*. This is a conical motion due to the misalignment of the axis of maximum moment of inertia and the angular momentum vector. The maximum moment axis of the vehicle essentially rolls around in a cone around the angular momentum vector. Nutation is usually a very transient motion, since it is eliminated by energy dissipation. Indeed, many spin-stabilized spacecraft are equipped with nutation dampers specifically to reduce the time spent in this state.

A third term used in spacecraft attitude determination is *coning*. Coning refers to the apparent conical motion indicated by a sensor which is not aligned with a principal axis. Even in a perfectly steady rotation, a misaligned sensor will appear to indicate motion in another axis.

A torque can be impulsive, i.e., of short duration. In such a case, the torque–time product yields an angular momentum increment which changes the direction of the vehicle's angular momentum before the body itself has had time to move accordingly. In this situation, the body spin axis (usually the axis of maximum moment of inertia) will be misaligned with the angular momentum vector. The vehicle will appear to wobble—this is the nutation motion mentioned earlier. The amplitude of this wobble will decrease with time as energy dissipation realigns the spin axis with the new angular momentum vector. The rate of the wobble depends on the moments of inertia of the object: for a flat disk, where the axial moment of inertia is exactly double the transverse moments of inertia, the wobble period is half of the spin period.

Aerodynamic Forces and Torques

Angular dynamics of aircraft are usually described by three motions: roll, pitch, and yaw. *Roll* denotes motion about a forward direction. *Yaw* is motion about a vertical axis, while *pitch* is motion in a plane containing the vertical and forward directions.

Aerodynamic forces and moments can be considered in several ways. Ultimately, all forces must be expressed through pressure normal to, and friction along, the surfaces of the vehicle. For most of the flows considered here, viscous forces are modest and only the pressures are significant. For a body to generate lift, pressures on its upper surfaces must in general be lower than those on its lower surface.

Another perspective is that the exertion of force on the flying object must manifest itself as an equal and opposite rate of change of momentum in the airstream. If an object is developing lift, it must therefore push the air down. Streamlines must therefore be tilted downwards by the object.

The distribution of pressures on the flying object will yield a resultant force that appears to act at an arbitrary position, the center of pressure. No torques about this point are generated.

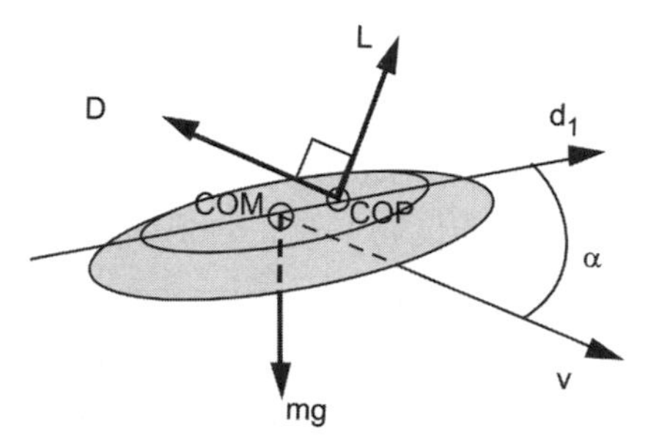

Figure 1.3. The weight (*mg*) of the vehicle acts at the center of mass (COM) whereas the aerodynamic force acts at the center of pressure (COP) and is usually defined by a lift and drag, orthogonal and parallel to the velocity of the vehicle relative to the air. Because the COM and COP are not in the same place, there is a resultant pitch torque. The airflow hits the vehicle at an angle of attack α. Figure by Sarah Hummel, used with permission.

More usefully, the force is calculated as if it acted at the geometric center of the vehicle. Usually this force is expressed in three directions, referred to the direction of flight. *Drag* is along the negative direction of flight; *lift* is orthogonal to drag in the vehicle-referenced plane that is nominally upwards. The orthogonal triad is completed by a side-force. The forces are supplemented by a set of moments (roll, pitch, and yaw). These determine the stability of a vehicle in flight.

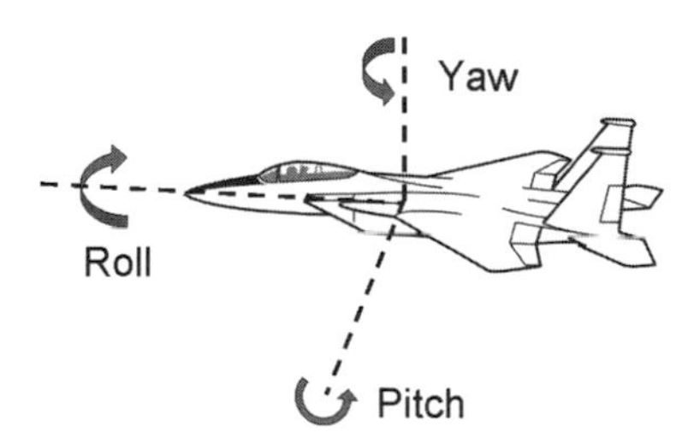

Figure 1.4. Roll, pitch, and yaw.

Both forces and moments are normalized by dimensions to allow ready comparison of different sizes and shapes of vehicles. The normalization for forces is by the dynamic pressure ($0.5\rho V^2$) and a reference area (usually the wing planform area). Dividing the force by these quantities, the residual is a force coefficient. These coefficients—dimensionless numbers usually with values of 0.001 to 2.0—are typically functions of Mach and Reynolds number (which are generally small and constant enough, respectively, to be considered invariant in the applications here) and of attitude.

The attitude (the orientation of the body axes in an external frame—e.g., up, north, east) may be compared with the velocity vector in that same frame to yield, in still air, the relative wind, i.e., the velocity of the air relative to the vehicle. In cases where there is a nonzero wind relative to the ground, an ambient wind vector may be added to the relative wind. In addition to changing the speed of the air relative to the vehicle, wind may be instrumental in changing the angle of the airflow relative to the body datum.

The most significant angle is that between the freestream and the body datum in the pitch plane. This is termed the *angle of attack*, and it is upon this parameter that most aerodynamic properties such as lift and moment coefficients display their most significant sensitivity.

A second angle is relevant for conventional aircraft, and this is the angle of the freestream relative to body datum in the yaw plane: this is the *sideslip angle*, but this is not significant for the cases described in this book—for the most part if a sideslip angle were to be defined relative to a body datum, it would vary rapidly owing to the body spin.

The spin rate is usually not itself of intrinsic aerodynamic interest. However, when multiplied by a body length scale (the span of a boomerang, or the radius of a Frisbee) it corresponds to a tip or rim speed. This speed can be significant compared with the translational speed of the body's center of mass, and thus a measure of the relative speed is used, referred to as the *advance ratio*. Note that definitions of this term vary, but here we use $\omega r/V$.

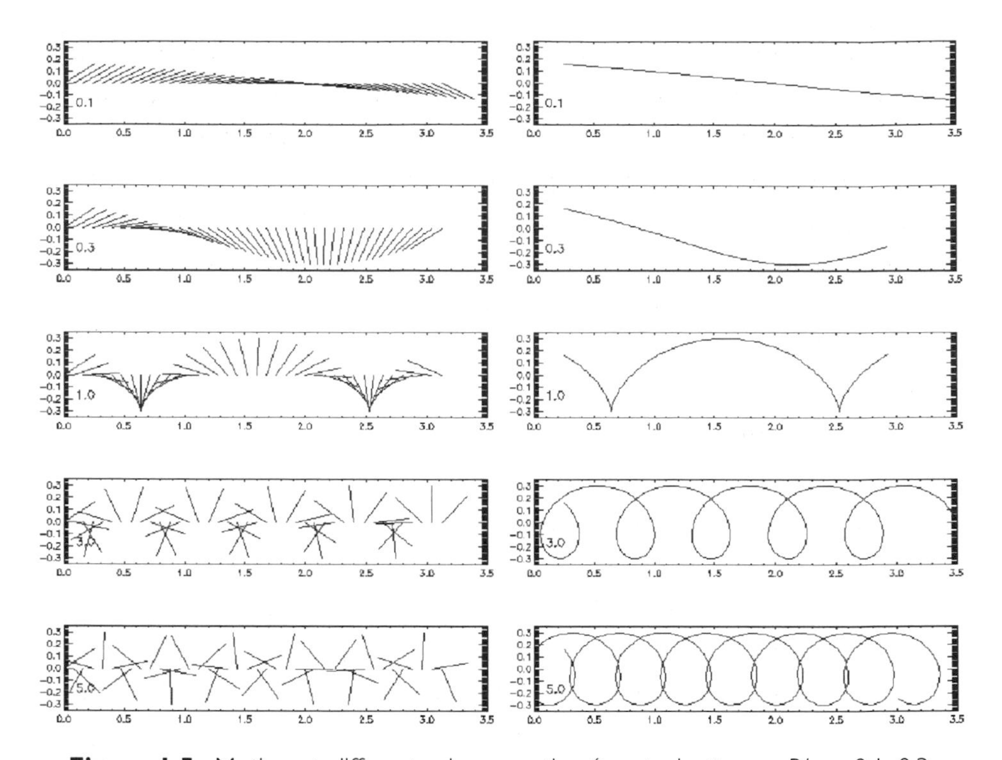

Figure 1.5. Motion at different advance ratios (top to bottom, $\omega R/v$ = 0.1, 0.3, 1.0, 3.0, 5.0). On the left is shown a sequence of positions of a stick thrown at this advance ratio—a stroboscopic photograph would show this pattern. On the right is the trace that is made by the end of the stick—if a lamp were attached to the tip of the stick and the stick were thrown at night, a camera with the shutter held open would record this trace. The traces are cycloidal.

Drag and the Dimensionless Parameters of Flight

Several parameters describe flight conditions—the density and viscosity of the fluid, the size and speed of the object, and so on. Most usefully, these properties are expressed as dimensionless numbers to indicate the ratio of different forces or scales. Because flow behaviors can be reproduced under different conditions but with the same dimensionless numbers, these numbers are often termed similarity parameters.

The most familiar of these may be the *Mach number*. This is simply the ratio of the flight speed to the speed of sound in the medium. Since the sound speed is the rate of propagation of pressure disturbances (= information), a Mach number in excess of 1 indicates a supersonic situation where the upstream fluid is unaware of the imminent arrival of the flying object, and the flow characteristics are very different from subsonic conditions. In particular, a shock wave forms across which there is a discontinuous jump in pressure and temperature as the flow is decelerated. This shock wave typically forms a triangle or (Mach) cone around the vehicle, with a half angle equal to the arctangent of the Mach number.

A similar phenomenon can be seen in objects moving across the surface of a liquid—a series of surface waves propagate radially outwards, but the forward motion of the source convolves these circles within a triangular envelope whose apex becomes progressively sharper as the object moves more quickly. The objects discussed here for the most part are firmly subsonic and Mach number variations ($\ll 1$) are not of concern.

Much more important is the *Reynolds number*, the ratio of viscous to inertial (dynamic pressure) forces. This may be written $Re = vl\rho/\mu$, where v is the flight speed, l a characteristic dimension (usually diameter, or perhaps a wing chord) μ the (dynamic) viscosity[2] of the fluid, and ρ the density of the fluid.

---

[2] Note that there are two "types" of viscosity. Here we refer exclusively to dynamic viscosity, the ratio of the shear stress to the velocity gradient in a fluid. This is a constant for a given fluid, and is what is measured directly. The symbol for this property is usually Greek mu, and the units are those given by stress/velocity gradient = Newtons per meter squared, divided by meters per second per meter, thus Newtons per meter squared, times second, or Pascal-seconds (Pa-s) in SI units. Another somewhat archaic unit sometimes used is the Poise (P). 1 P = 10 Pa-s.

For reference, though we do not use it in this book, the "other" viscosity is kinematic viscosity, which is dynamic viscosity divided by density. Make sure you use the right one!

The effects of Reynolds number may be most familiar in the variation of drag coefficient. At very low values of *Re* (< 1), viscous forces dominate, and the drag coefficient for a sphere is equal to 24/*Re*, which may be very large. Since the drag coefficient C_∂ is defined with respect to the fluid density, which directly relates to the (insignificant) inertial forces, this relation for C_∂ is equivalent to substituting a formula for drag that ignores density and instead relates the drag only to viscosity. This substitution leads to Stokes' law for the fall velocity of spheres in a viscous fluid.

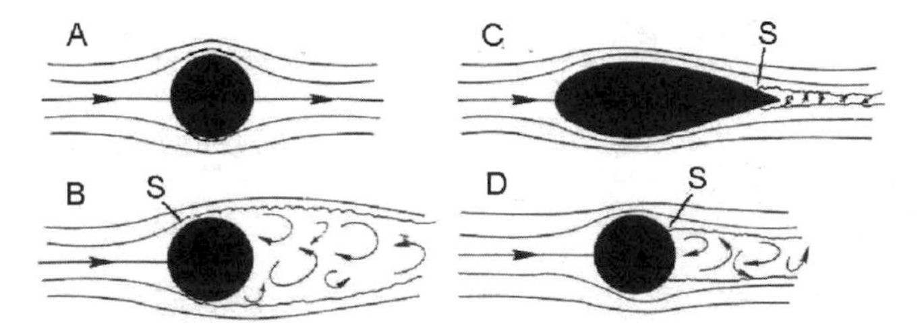

Figure 1.6. Flow past a round object. In (A) the Reynolds number is low (< 10), and the flow remains attached throughout—pressure drag is tiny, but there is much viscous drag: the drag coefficient will be around (24/Re). In case (B), typical of Reynolds numbers of 1000–10^4, the flow separates at position S and the wake is wide—the drag coefficient C_d is therefore relatively large at about 1.2. In case (C) also at $Re \sim 10^4$, a streamlined shape permits the boundary layer to remain attached for longer, so S is close to the trailing edge and the wake is narrow and drag is small: $C_d \sim 0.12$. In case (D), $Re \sim 10^5$, so the flow is more strongly turbulent (a similar effect can arise from roughening the surface with dimples, or by attaching a wire or seam near the front); it can remain attached and so drag is lower ($C_d \sim 0.6$) than for case (B).

As the Reynolds number increases (the flow becomes "faster"), the inertial forces due to the mass density of the fluid play a bigger and bigger role (Figure 1.6 (C),(D)). The flow becomes unable to stick to the back of the sphere, and separates. At first (*Re* of a few tens) the flow separates from alternate sides, forming two lines of contrarotating vortices. This is sometimes called a von Karman vortex street, although (Tokaty, 1994) it had been observed long before von Karman. This alternating vortex shedding is responsible for some periodic flow-driven phenomena such as the singing of telephone wires, and the vortices are sometimes seen in geophysical fluid flows (Figure 1.7).

Control of drag is essentially equivalent to controlling the wake—whatever momentum is abstracted from the flow onto the object (or vice versa) is manifested in the velocity difference between the wake and the undisturbed fluid. If the wake is made more narrow, then the momentum dumped into it, and thus the drag, will be kept small. One way of doing this is by streamlining (Figure 1.6, (C)). Another circumstance is to make the boundary layer (the flow immediately adjacent to the object) turbulent, which allows it to "stick" better to the object and thus make the wake more narrow (Figure 1.6, (D)).

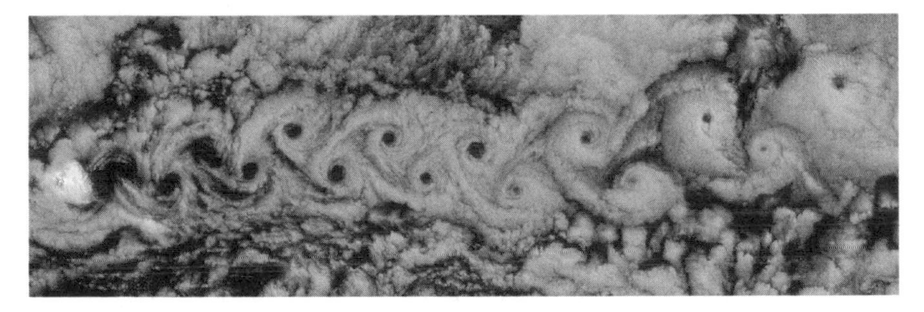

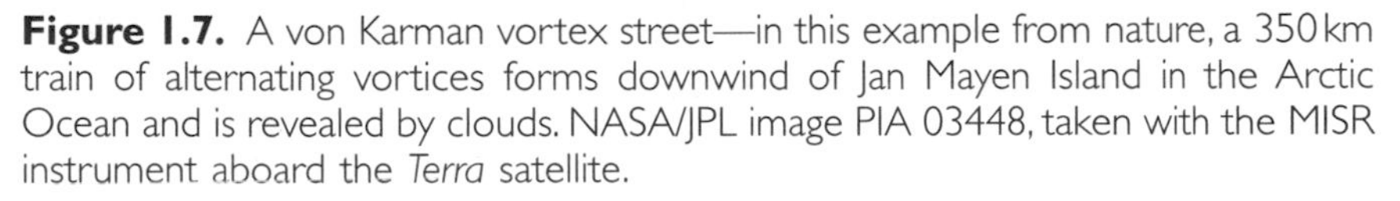

Figure 1.7. A von Karman vortex street—in this example from nature, a 350 km train of alternating vortices forms downwind of Jan Mayen Island in the Arctic Ocean and is revealed by clouds. NASA/JPL image PIA 03448, taken with the MISR instrument aboard the *Terra* satellite.

As we discuss in the next chapter, symmetric control of the boundary layer is of course known in golf, whereby the pimples on the ball increase the surface roughness so as to make the boundary layer everywhere turbulent. The turbulent boundary layer is better able to resist the adverse pressure gradient on the lee (downstream) side of the ball and remains attached longer than would the laminar layer on a smooth ball. The result is that the wake is narrower and so drag is lessened.

Similar boundary layer control is sometimes encountered on other (usually cylindrical) structures that encounter flow at similar Reynolds numbers. An example is the bottom bar on a hang glider. This bar can be faired with an aerofoil, but such a shape is harder to grip with the hands, and a cylindrical tube is rather cheaper. However, a smooth cylinder has a high drag coefficient, so a "turbulating" wire is often attached at the leading side to trip the boundary layer into turbulence and so reduce drag.

Another similarity number is the Knusden number K_n, which is relevant only for flight at extremely high altitudes. K_n is the ratio between the vehicle dimension and the mean free path of the fluid molecules—in familiar settings fluids behave, well, like fluids and the Knusden number is large. However, if $K_n < 1$, as encountered by satellites and entry vehicles, the gas molecules act as billiard balls without interacting with one another. In this setting the drag coefficient is invariably around 2; the flow basically dumps all of its momentum onto the vehicle and slides off, with a near-perfect vacuum left on the lee side. In fact, this is how Isaac Newton originally visualized fluid flow.

Lift

It is often said that the airflow across the curved top of a wing is faster than across the bottom, and since Bernoulli's theorem states that the sum of the static pressure and dynamic pressure in a flow are constant, then the faster-flowing (higher dynamic pressure) air on the upper surface must "suck" the wing upwards. This is sometimes true in a sense, but it is a rather misleading description—it fails utterly, for example, to explain why a flat or cambered plate can develop lift. In these cases in particular, cases which are approximated in many situations explored in this book, the air travels the same distance over top and bottom, and so faster flowing air is not required on the upper surface, at least not from geometric considerations alone.

It is better to take a step back. Conservation of momentum dictates that if the airflow is to exert a lift force on the wing, then the wing must exert a downward force on the air. The flow of air past the wing must be diverted downwards. Whatever causes the lift, a result will be a downwards component of velocity imparted to the air. In an idealized sense, one can imagine the downward diversion of the airflow as a rotation of the streamlines, and consideration of the wing as a circulation-inducing device is a powerful idea in fluid mechanics (e.g., Tokaty, 1994).

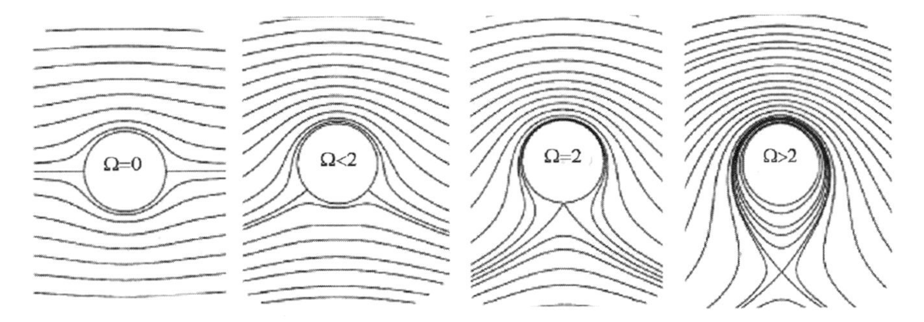

Figure 1.8. An idealized flowfield—so ideal, in fact, that the flow can be in either direction—about a rotating cylinder at an advance ratio **Ω**. Seen from afar, the deviation of the streamlines is just the same as that caused by a lifting wing.

As for pressure, which is force per unit area, there will indeed be on average a lower pressure on the upper surface of a lifting wing than on the lower surface. The net force is simply the integral of the pressure over the wing. Whether you consider the velocity field of the flow the cause or the effect of the pressure distribution is not important.

Indeed, since both of these aspects of the flow will depend on the shape of the wing, its orientation (a flat plate inclined slightly upwards to the flow will obviously divert the flow downwards), and on how the flow stays attached to the wing, one might consider both the flowfield and the pressure field to be effects. But it is not always true that the air flows faster across the upper surface.

The attachment of the flow is crucial. Once the flow has passed the suction peak—the area on the upper surface of the wing where pressure is least—the boundary layer may struggle to remain attached. If the layer separates, the drag will increase (as for a smooth sphere earlier) and the lift will be reduced. This condition is known as the "stall."

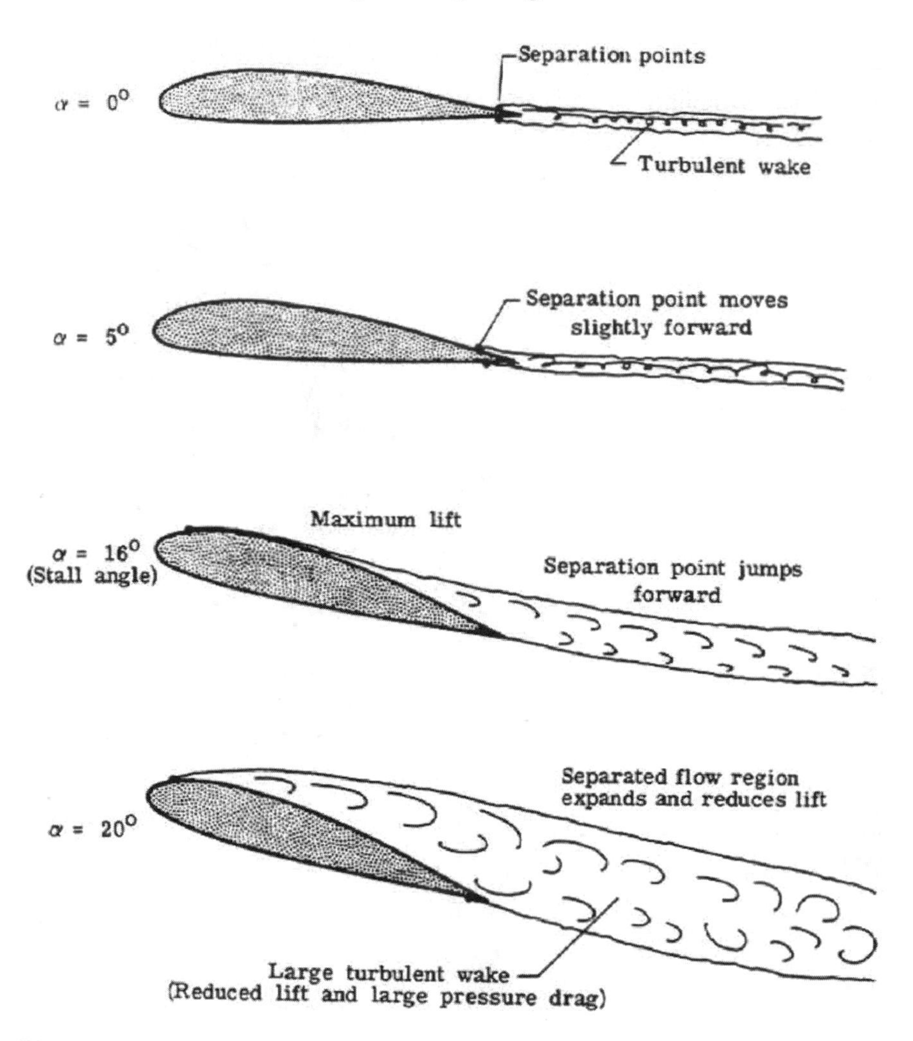

Figure 1.9. Onset of stall as the angle of attack increases. From NASA SP-367.

Robins–Magnus Effect

An effect that usually receives only modest treatment in texts on aerodynamics is one that is central to the theme of this book. This is the role of spin in aerodynamics. As is well-known to sportsmen everywhere, and as we discuss in detail in the next chapter, even a perfectly symmetric sphere can be made to veer in flight by causing it to spin.

This force is often referred to as the Magnus effect, after the German physicist Gustav Magnus, who demonstrated it in his Berlin laboratory in the middle of the nineteenth century. However, the effect was first studied systematically by the English gunnery expert Benjamin Robins in the eighteenth century as an explanation for the scatter of spherical shot. In essence, the airflow over the side of a ball (or cylinder) that is spinning against the direction of fluid flow will cause the flow to separate earlier, while the flow running with the rotating surface will stick longer. The result is that the wake is diverted sideways. Because the flow is diverted sideways, there must be a reaction on the spinning object—a side-force.

This side-force causes spinning objects to veer in flight—a ball "rolling forward" (i.e., with topspin) will tend to swing down, while a ball with backspin will tend to be lofted upwards—a key effect in golf. In essence the ball "follows its nose." If the spin axis is vertical, then the ball is diverted sideways.

Note that some books give an explanation of the Robins–Magnus force as the relative airspeed on the oncoming side of the sphere being higher and thus the dynamic pressure being higher. This is a rather inaccurate and misleading explanation (indeed, it might be construed as indicating the wrong direction for the force). Think of the Robins–Magnus force just as lift and drag—momentum dumped into the wake via control of flow separation. In the next chapter we will discuss other ways of controlling flow separation and thereby affecting dynamics of flying objects.

No discussion of the Robins–Magnus force would be complete without mentioning the Flettner rotorship. Anton Flettner, a German, had the idea that spinning cylinders on vertical axes would act like wings. In essence the cylinders were sails, but sails that could be more easily controlled than large flapping pieces of cloth. Indeed, an ocean-going ship was built in 1924 (the *Buckau*, renamed the *Baden Baden* in 1926) with two cylinders spun by electrical motors, and it seems to have performed very well. It was able to tack into the wind at about 20 to 30°. However, despite sailing to New York and South America, it was still just a sailing ship and thus could not move in a dead calm.

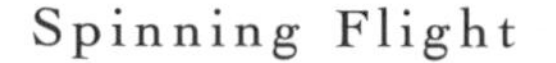

Figure 1.10. Deviation of flow streamlines downwards by (implying a lift force upwards on) applying spin to a 4.5 inch cylinder. These color-inverted smoke-stream photos in an early wind tunnel show the Robins–Magnus effect. From NACA TN 228.

References

Magnus, G., *Ann. Phys. Chem.* 88, 1–14, 1953.

Rizzo, F. The Flettner rotor ship in the light of the Kutta-Joukowski theory and of experimental results, *NACA TN* 228, Oct. 1925.

Robins, B., *New Principles of Gunnery Containing the Determination of the Force of Gunpowder and Investigation of the Difference in the Resisting Power of the Air to Swift and Slow Motion*, 1742.

Talay, T., Introduction to the aerodynamics of flight, *NASA SP-367*, 1975.

Tapan K. Sengupta and Srikanth B. Talla, Robins–Magnus effect: A continuing saga, *Current Science* Vol. 86, No. 7, 10 April 2004.

Tokaty, G. A., *A History and Philosophy of Fluid Mechanics*, Dover, 1994.

2
Aerodynamics of Balls

Introduction and History

In this chapter we consider balls and their flight mechanics. Some balls have peculiar dynamics owing to nonspherical shape, whereas others are round but exploit various aspects of boundary layer control. These aspects include overall surface roughness, such as the dimples of a golf ball, and asymmetric roughness, such as the seam of a cricket ball or baseball. Still other complications are introduced by spin, which may generate lift via the Robins–Magnus effect. The additional complications of bouncing and the generation of spin by impact will be discussed in a later chapter.

Many flying objects are chosen to be smooth spheres for a variety of reasons. First is manufacturing. Musket balls and other shot are round for ease of manufacture. Indeed, lead shot represents perhaps the first example of microgravity materials processing; molten lead is poured through holes and allowed to fall some tens of meters into water. The fall time is long enough to permit surface tension to draw the lead blobs into spheres, which are then quenched solid by the water. In addition, the round shape in the absence of spin leads to zero transverse aerodynamic forces and thus predictable flight, as for the stone balls used in cannon and siege catapults and ballista before the era of the musket.

Similarly, many balls used in games such as table tennis are smooth and deliberately symmetrical to avoid anomalous behavior. However, the balls used in several sports are deliberately rough, and usually rough in a nonuniform way.

Isaac Newton in 1672 had observed that balls in the game of tennis could be affected by spin: "For, a circular as well as a progressive motion . . . its parts on that side, where the motions conspire, must press and beat the contiguous air more violently than on the other, and there excite a reluctancy and reaction of the air proportionately greater."

The effect of spin on cannon shot was measured quantitatively by the English military engineer Benjamin Robins in 1742. Robins was a very practical artilleryman, and had observed that the scatter in the fall of shot of otherwise reproducible cannon varied not in direct proportion to the range (as it would if the shot left the cannon with some scatter in angle of flight) but rather more than that. Robins reasoned correctly that something—i.e., a force—was happening to the shot *after* it left the gun barrel. Since he was careful to exclude wind effects, and he could recover the shot afterwards to verify that the balls were still round, only spin was left as a possible cause for the deviations.

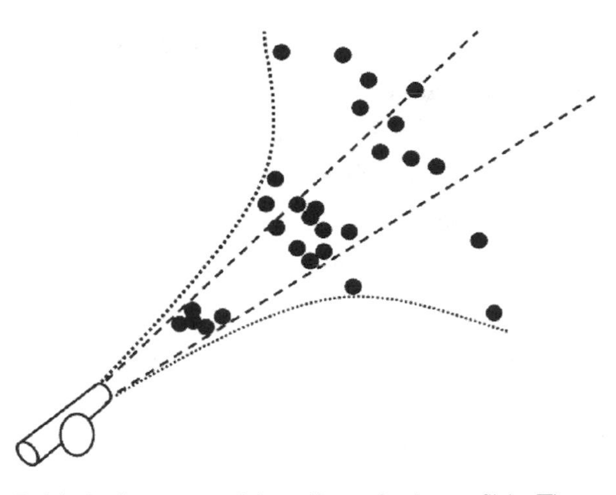

Figure 2.1. Robins's discovery of the effect of spin on flight. The scatter of controlled shots varied not in proportion to the distance (bounded by the straight, dashed lines) as expected from launch errors but rather grew more quickly with distance, implying they were accelerated sideways (dotted lines) after leaving the gun.

Despite this methodical experimental approach, Robins was not universally believed. The great Swiss mathematician Leonhard Euler, even though he was so enthused by Robins's work that he himself translated it for publication in German in 1745, dismissed Robins' observation of side-force due to spin. It was inconsistent with Euler's own idealized theories of fluid motion. Euler's reputation was such that his skepticism of Robins' observation was enough to suppress the idea of spin-induced side-forces.

The force was "rediscovered" by the talented German physicist and educator Gustav Magnus, whose work was popularized in England by Lord Rayleigh. Because the distinguished Rayleigh attributed the side-force to Magnus, most literature on the topic refers to the "Magnus force." Appropriately for this chapter, however, Rayleigh, among his much more important work, published a short paper in 1878 on the dynamics of tennis balls.

The systematic study of ball aerodynamics relies, in part, on the uniformity of the product at hand. One could not, for example, make a careful study of the aerodynamics of sticks of wood if one were limited to whatever tree branches were at hand. Similarly, there would have been considerable variation between tennis balls in centuries before the 19th century, since they were hand-made items, varying from craftsman to craftsman. The same is true for other sporting goods. Perhaps Robins was ahead of his time in that only the exacting requirements of precision military equipment furnished sufficiently uniform apparatus with which to conduct useful comparative experiments.

It is of some academic interest that most of the published work on ball aerodynamics is published by physics teachers: balls are familiar and easily procured items on which to conduct experiments. Sporting goods manufacturers have large quantities of data on the properties of their products, but most of this is proprietary. Not only are sporting goods a major business, but the role of technology in sports has increased substantially in recent decades with the advent of television. Associating sports success with a product means an enhanced image, and therefore increased sales. If a performance advantage can be gained by using a different ball or club, then it pays to understand exactly how well it works and why, but not necessarily to share that information with one's competitors.

In this chapter we will consider the dynamic and aerodynamic aspects of a number of ball sports. Before considering various sports in turn, we will first consider the characteristics that affect a ball's flight.

Balls: Types and Characteristics

For the most part, the size range of balls used in sports is limited to the range that can be conveniently manipulated and thrown with a single hand. This spans the range from a couple of centimeters diameter (squash balls, golf balls, etc.), where the ball is placed or thrown into play single-handedly, to larger balls (~ 25 cm) such as footballs and basketballs which can—with skill—be thrown from a single open hand, but are often thrown with two.

The speed range is in general that which can be achieved by the unaided human arm—a few tens of meters per second, with the peak speed falling with ball mass once the ball adds significant inertia to the arm–ball system. Particular fast-movers are cricket balls and baseballs, which at ~150 g are perhaps an optimum mass for throwing. Golf balls fly particularly fast because of the large moment arm of the club used to strike them, and the elastic properties of club and ball. Footballs (soccer balls) generally fly faster than other balls of the same size (such as basketballs) because they are kicked—the long moment arm of leg, and the large mass of the foot, make an efficient club.

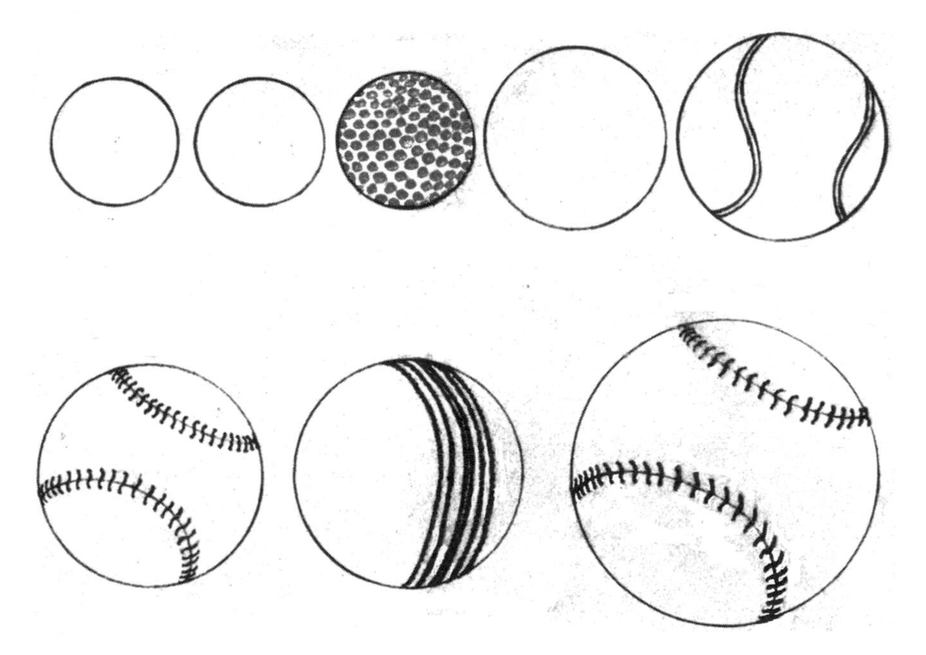

Figure 2.2. Drawing (to scale) of "small" balls. Left to right (top): table-tennis, squash, golf, racquetball, tennis. Left to right (bottom): baseball, cricket, softball.

The key property affecting a ball's general flight is its mass:area ratio. Squash balls and table-tennis balls have the same size, but the latter has a much lower mass. In principle, one could throw a table-tennis ball a little faster than a squash ball, but one would be unlikely to throw it further, since air drag has a much stronger effect on the table-tennis ball. At a given speed, the drag force on the two balls is the same, but the change in flight speed (i.e., the acceleration, equaling force/mass) is much higher for the table-tennis ball.

The other aspect affecting a ball's drag (without spin or asymmetric boundary layer control, a ball has no lift) is the flight Reynolds number. The drag coefficient of spheres is somewhat constant over a wide range of Reynolds number, but falls appreciably at a critical Reynolds number which ranges between about 4×10^4 to 4×10^5, depending on surface roughness. This drop is illustrated in Figure 2.3 below. It can be seen that the golf ball, which has a relatively high surface roughness, has a drag coefficient that drops at a low Reynolds number. The somewhat smoother soccer ball has a drop at a higher Reynolds number, but because the ball itself has a larger diameter, this critical Reynolds number corresponds to a lower flight speed than the golf ball. A perfectly smooth sphere has the highest critical Reynolds number.

Table 2.1. General properties of sports balls.

Note that most ball specifications in the UK or U.S. are given as a range, and specified in Imperial ("English") units. I have given here a representative metric figure.

Ball	Diameter (cm)	Mass (kg)	Mass/Area (kg/cm^2)	Speed (m/s)	Reynolds Number
Table Tennis	4	0.0027	0.0002	25	5.9E+04
Squash	4	0.024	0.0019	60	1.4E+05
Racquetball	5.7	0.0418	0.0016	40	1.3E+05
Golf	4.2	0.045	0.0032	70	1.7E+05
Tennis	6.5	0.057	0.0018	45	1.7E+05
Baseball	7	0.141	0.0037	40	1.7E+05
Cricket	7	0.163	0.0042	30	1.2E+05
Softball	9.7	0.185	0.0025	10	5.7E+04
Volleyball	21	0.284	0.0008	15	1.9E+05
Soccer	21	0.429	0.0012	20	2.5E+05
Football (American)	22	0.43	0.0008	15	2.3E+05
Basketball	24	0.625	0.0013	8	1.2E+05
Shot Put (mens')	12	7.5	0.0663	14	9.9E+04

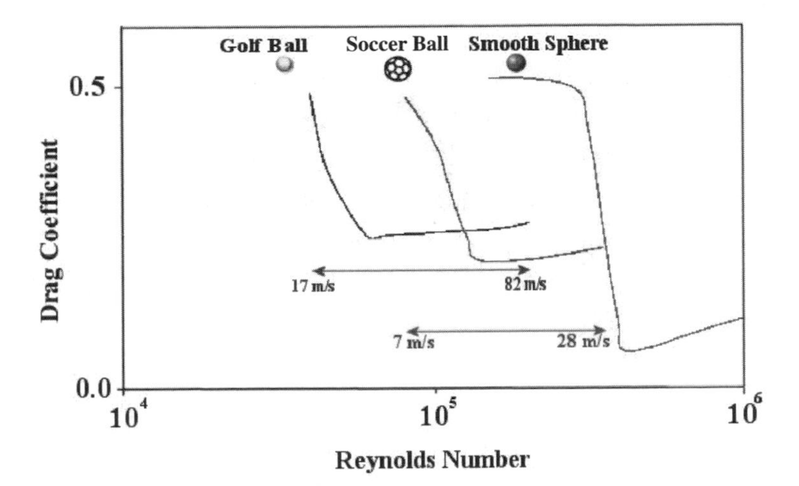

Figure 2.3. Drag coefficient of golf ball and soccer ball as a function of Reynolds number. The corresponding flight speeds are shown. Lower surface roughness gives a higher critical Reynolds number at which the drag drops by a factor of 2 or more.

Tennis

We will begin our discussion of individual sports in a vaguely chronological manner. Tennis is perhaps an older game than most of the others discussed—hence the early observation of the nonballistic trajectories followed by balls. None other than Isaac Netwon, at the time 23 years old, observed that tennis ball flight was affected by spin. Lord Rayleigh (William Strutt, perhaps most famous for his analytic work on thermally driven convection) also considered the swerving motion of a tennis ball.

The pattern of stitching on a tennis ball is more or less the same as that on a baseball: two patches of hide are sewn together in a yin-yang pattern; the convex parts of the broadly figure-eight shape of each half match up, after accommodating the three-dimensional curvature of the ball, with the concave part of the other half. From the first Wimbledon tennis tournament in 1902 until around 1929, the hand stitching meant that no two balls behaved exactly alike. Then a new vulcanizing process was introduced to manufacture the rubber core and bond the wool cloth to the outside—bonding by hot press eliminated the need for stitches, and a special refrigerated container was developed about a decade later. Original balls were white; yellow balls for improved visibility were introduced in 1986.

The texture of a tennis racquet is such that the effective coefficient of friction between the racquet and ball is quite large, permitting very large spin rates to be induced; indeed, regulations exist on the stringing style of a racquet (see following chapter) to restrict the amount of spin.

The spin permits large side-forces (i.e., "lift") to be developed by the Magnus force. A tennis ball flies at speeds well above the critical Reynolds number, and thus the boundary layer is tripped into turbulence with a spin-dependent location. The usual application is in the "topspin lob" whereby a topspin causes the ball to dive downwards, permitting a fast shot that still hits the court within the permitted boundary.

A subtle effect can occur with tennis balls, which have a fuzzy felt surface, that is not exhibited by leather or plastic balls. The fibers that make the ball fuzzy are not rigid, and thus a strong airflow will make them lie flat. When they lie flat, they intercept less of the airflow, and thus experience less drag than they would if they had remained upright. Thus, the drag coefficient of a tennis ball will be lower than might be expected at high speeds, all else being equal. (Of course, this small effect is superimposed on the more general variation of drag coefficient with Reynolds number.)

The International Tennis Federation is considering the introduction of a larger ball. This, by causing higher drag, would lead to a slower game. Modern racquets and players hit the ball so fast that serves tend to dominate the game, making it less exciting to watch.

Golf

The game of golf involves some of the highest speed motion in ball sport—a small ball is struck with a massive club, which provides the largest possible moment arm, the length of the lever (arm + club) being dictated by the distance between the shoulder and the ground.

The Scottish physicist Peter Guthrie Tait, while not a widely encountered name in modern physics, was nonetheless a star of his time. He was a correspondent of James Clerk Maxwell and co-wrote a physics textbook with Lord Kelvin (indeed, it was in a letter to Tait that Maxwell suggested the idea of a "light-fingered being" able to sort out individual molecules—"Maxwell's Demon"). He was an enthusiastic promoter of quaternions, a 4-dimensional representation of angles which avoided certain singularities, a representation that is used widely in spacecraft attitude dynamics.

In a series of papers in the late 1890s Tait, an avid golf player, attempted to establish the physical and mathematical basis for the observed characteristics of the game of golf, and in particular the non-ballistic behavior of golf balls.

Tait used a pair of whirling paper discs punctured by a stiff wire to estimate the velocity of the golf club to which the wire was attached. By comparing the position of the holes in the two discs (mounted on the same axle) as punctured by the wire, the speed of the wire could be calculated from the rotation speed of the discs, their separation. Although he acknowledged that his test subject may have been distracted by the apparatus, and thus not playing his best, Tait determined that the club head was moving at between 60 and 100 m/s when it struck the ball, and the ball moved rather more than this. Furthermore, Tait showed, by means of a thin paper tape attached to the ball, that the ball was spinning rapidly after it was hit by the club. He estimated that the contact time between the golf ball and the club was only about one ten-thousandth of a second.

Tait attempted to calculate algebraically the trajectory of golf balls acting under the influence of drag, and noted the law of diminishing returns—that a progressively faster ball does not go that much further—speeds of (100, 200, . . . , 600) feet per second would lead to ranges of (112, 277, 400, 497, 571, 600) feet. But the main puzzle was why a ball could spend so long in the air, and go so high. Spin had to be the key.

By suspending a golf ball by a fine wire which could be rotated rapidly with a reduction gear (like that on a hand drill), and at the same time swinging the ball on the wire like a pendulum, Tait showed that the spin caused large side-forces to be developed (he credits Robins with the idea). These experiments supported Tait's calculations that a golf ball's flight lasted far longer than would be possible if the ball did not develop lift.

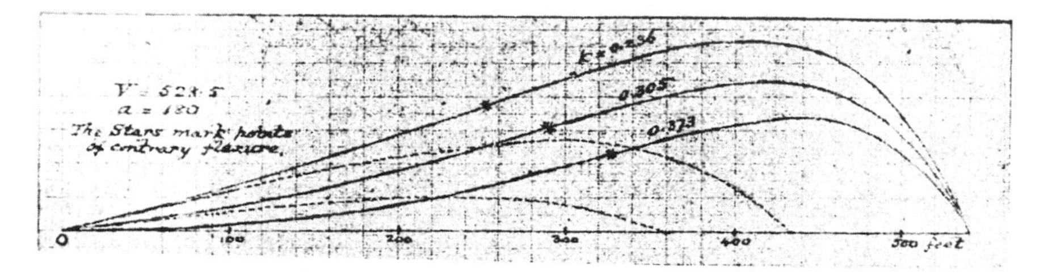

Figure 2.4. Golf ball trajectories computed by Tait.

The shape of a gold club is designed not only to provide a translational velocity impulse, but also to impart a backspin to provide lift. The angle of the club head, its shape, and the presence of grooves to increase friction all play a role here.

The original golf balls were made of wood. By the early 17th century handmade balls made of cowhide stuffed with feathers were in use. But it was gutta percha (dried sap from the rubber tree) that made the game more popular, the balls being called "Gutties."

It was soon realized that an old, roughened ball appeared to travel further than a smooth one. This of course is due to the roughness-induced transition of the boundary layer into turbulence, which allows the wake to be narrower and thus drag to be reduced.

The deliberate roughening of a golf ball to improve its flying characteristics began circa 1880—initially with bumps rather than dimples, one popular pattern being called a bramble. Multiple-layered balls (with a rubber core surrounded by a wound rubber thread and a gutta percha skin) were mass-produced soon after the start of the twentieth century, and the familiar dimpled pattern was introduced by about 1930. The conventional design in the UK and U.S. respectively had 330 or 336 dimples.

Wind tunnel tests by Bearman and Harvey in 1949 using a motorized assembly to spin the ball show that the classic golf ball, at a Reynolds number of 10^5, has a lift coefficient that varies from 0 with zero spin rate to about 0.3 when the advance ratio (the circumferential speed divided by the flight speed) reaches about 0.4, corresponding to about 6000 rpm. The drag coefficient rises from about 0.27 to 0.34 over the same range of advance ratio. The lift coefficient for a given spin rate is highest at low speeds, since the advance ratio is correspondingly high, although the variation is small above speeds of 55 m/s.

Populating the surface of a sphere with a regular pattern without a preferred orientation is not a trivial problem, and thus not only the number or density of the dimples, but also their arrangement on the ball surface influences the ball's performance in play.

The Wilson sporting goods company introduced the Ultra 500 ball in 1995. As its name suggests, it has 500 dimples, arranged to form 60 triangular faces. There are in fact three sizes of dimple, arranged to maximize lift while minimizing drag. The pattern was designed by an aerodynamicist, Bob Thurman, who formerly worked for Martin Marietta on the loads and dynamics of the space shuttle's external tank. Another new innovation recently developed by Procter and Gamble is to use polygonal dimples—these can be packed closer than circular ones.

Additional studies in ball aerodynamics are being made by computational fluid dynamics (CFD) methods, wherein the pressures and velocities of air in many tiny parcels of space are computed, and the transfer of momentum between them determined. Growth in computational hardware capabilities, and improved software tools to set up, run, and explore the results from these large calculations have made significant advances in recent years, and on many large aerospace projects (such as the Boeing 777 airliner, or the Deep Space 2 Mars Microprobes) are beginning to supplant wind tunnel testing.

Figure 2.5. Computational fluid dynamic simulations of the flow around a spinning golfball. The flow is the same in both cases: on the left the vectors show the flow relative to the ball; on the right is the flow relative to the undisturbed air, which more readily shows the perturbation caused by the ball, and in particular the curvature to the flow—and thus the lift—induced by the ball's spin. Images courtesy of Fluent, Inc.

The evolution of the dimple pattern has played a role in improving golf performance; between 1980 and 2004 the average driving distance on the PGA tour has increased by more than 10%.

Golf ball manufacturers are in a serious business—selling balls that can cost $40 or more apiece to tens of millions of golfers. The US Golf Association tests balls in a ballistic range facility where balls are struck down a 25 m long chamber. Balls that fly too well are not approved—after all, longer golf drives would eventually require longer golf courses!

It is interesting to note Tait's observation that the game of golf in its present form owes itself to the existence of the Earth's atmosphere. Were it not for the drag of the air, a golf ball would strike the ground at the same speed with which it was struck, some 500 feet per second as Tait puts it—"The golfer might deal death to victims whom he could not warn." He also observes that this illustrates the service of the atmosphere in protecting us by "converting into heat the tremendous energy of the innumerable fragments of comets and meteorites which assail the Earth from every side with planetary speeds."

Of course, Tait's vision of golf in vacuum was in fact demonstrated in an improvised form on the Moon by Apollo astronaut Alan Shepard.

Shepard was the Commander of *Apollo 14,* and towards the end of his 35-hour stay on the Moon, on February 6, 1971, he made one of the most famous golf shots in history. He had brought with him a golf ball, and attached a 6-iron head to the handle of a multipurpose sampling tool (for picking up moonrocks). His first attempt just pushed the ball into the dust. Another swing clipped the ball, sending it just a few feet to the side. His third attempt connected, sending the ball off on a low trajectory.

Shepard produced another ball, dropping it to the ground. After getting in position, he swung (not trivial in a pressurized space suit, which somewhat restricts movement) and made a solid connection with the ball, which stayed up (not up in the air!) for about 30 seconds, traveling a respectable 200 yards or so—"miles and miles and miles" was the quote from the mission transcript. Note that the trajectory did not benefit from lift, but relied solely on the low lunar gravity. One might note that lunar golfers might benefit from using a wedge or other steeply raked club—in the absence of aerodynamic lift, the maximum range is achieved for a launch elevation of 45 degrees, much steeper than is typical for terrestrial golf.

Football (Soccer)

As befits the object of a game popular all around the world, the characteristics of an English football are, broadly speaking, isotropic. The patches that are stitched together to make a football form a truncated icosahedron (i.e., one of the perfect solid shapes known to the ancient Greeks, made only of 20 isosceles triangles, but with some corners chopped off). The resultant shape is one with 60 corners, 12 pentagonal faces (often painted black), 20 hexagonal ones (white), and 90 edges.

Carre et al. (2002) used video measurements (240 frames per second) of the trajectory of a kicked football (A Mitre Ultimax, diameter 215 mm, mass 415 g) to infer lift and drag coefficients. Their drag coefficient for nonspinning balls is a surprisingly strong function of kick speed, varying from about 0.05 at 20 m/s to 0.35 at 30 m/s, increasing rather than decreasing with Reynolds number. Perhaps some deformation of the ball occurs in hard kicks.

The lift coefficient was close to zero for a nonspinning ball (as the broadly symmetric pattern of seams would suggest); however, as the spin rate incrased to 50–100 radians per second, the lift coefficient rose to about 0.25, and was more or less constant for higher spin rates. Here, of course, the lift is due to the Robins–Magnus effect.

Based on these data, they note that a football kicked at 18 m/s at an angle of 24 degrees from horizontal could fly at a range of 10 m anywhere between 1.2 and 3.2 m from the ground, purely by varying the spin rate at launch—topspin of course leading to lower altitude and backspin providing lift upwards.

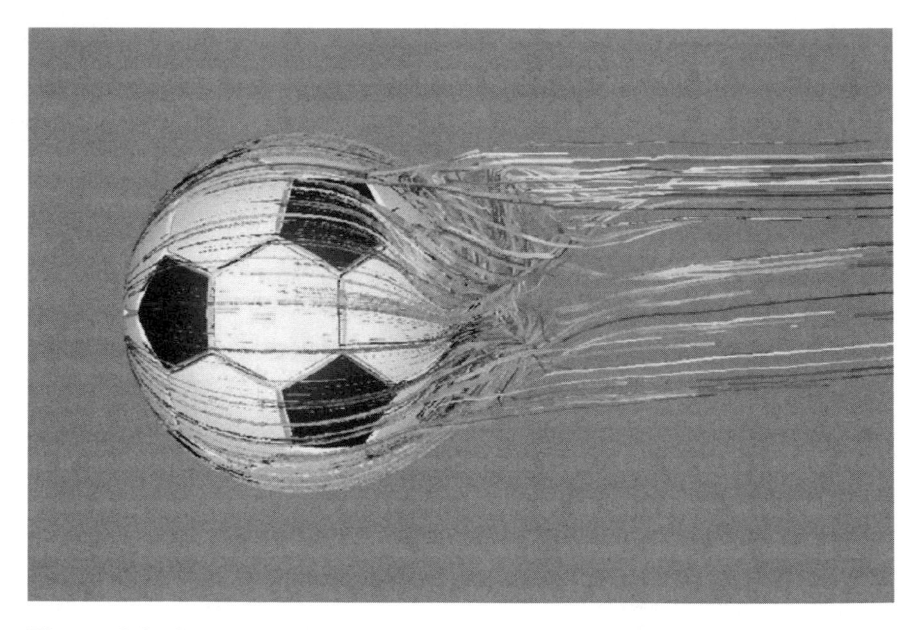

Figure 2.6. Streamlines on a nonrotating football. Notice the vertical near-wake structure. Image courtesy of Fluent, Inc.

The most common deliberate application of this side-force is in the corner kick, and in a free kick when a player attempts to bend the ball around a wall of defenders.

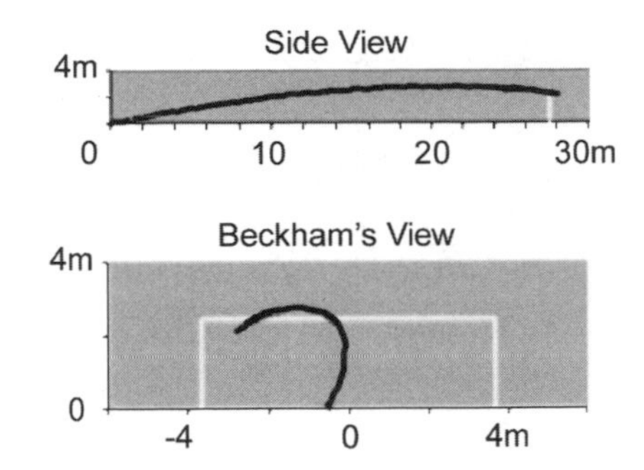

Figure 2.7. Trajectory (dimensions in m) of a free kick by England's David Beckham against Greece in October 2001. The ball looks as if it will overshoot the goal near its center, but breaks away to the left and sneaks under the bar.

CRICKET

A cricket ball has an equatorial plane with a set of stitching along it. This stitching acts as a boundary layer control structure—if the equator is held (by spin) at an angle to the airflow, then the flow on one half will encounter the stitches (and thereby have its boundary layer tripped into turbulence) on the leading hemisphere, while the other half will not encounter it until later.

The side that encounters the stitching may transition into turbulence, and thus is able to "stick" to the ball surface better through the adverse pressure gradient on the trailing side. In the absence of spin, the seam would slowly rotate around and average the side-force down to nothing. Thus, spin is applied to maintain a constant orientation of the seam, rather than to develop aerodynamic forces per se through the Magnus–Robins effect.

The optimum side-force (side-force coefficient CF ~ 0.3) is achieved when the seam equator makes an angle of about 20 degrees with the oncoming airflow. Similar results can be obtained with the seam at zero degrees if one hemisphere of the ball is smooth and the other rough. During play, bowlers allow one side of the ball to become rough, while rubbing (sometimes augmented with sweat or saliva) keeps one side smooth. Overt roughening is of course forbidden.

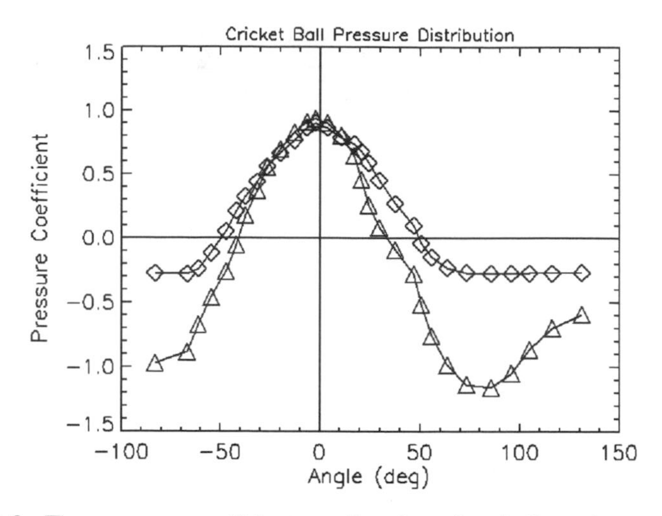

Figure 2.8. The pressure coefficient as a function of angle from the upwind seam, held at an angle of attack to the oncoming airflow of 20°. The diamonds correspond to a flight speed of 5 m/s, a subcritical *Re* of 2.5×10^4—the pressure distribution in this case is rather symmetrical and thus no side-force is developed. The triangles correspond to a faster bowl of some 25 m/s ($Re = 1.2 \times 10^5$)—the boundary layer is tripped on the upper side but not on the lower, leading to substantial lift.

An interesting change of behavior occurs at high speed. While a "swing" that amounts to almost 1 m in the "pitch" length of 20 m is possible at moderate speeds, such trajectories are only possible for a narrow range of throw speeds. The generation of a side-force depends on the differential separation of the boundary layer—one side must separate before the other. If the boundary layer on both sides were strongly turbulent, such that both separate at more or less the same place, then the seam would make little difference. This situation occurs in fast bowls—the ball can be thrown at up to 40 m/s ($Re \sim 1.9 \times 10^5$). Because the boundary layer becomes naturally turbulent even in the absence of the seam at $Re \sim 1.5 \times 10^5$, the side-force coefficient begins to fall off at this speed (~ 30 m/s).

Baseball

Baseball is a game which seems to have attracted a considerable body of at least casual scientific interest (e.g., Bahill et al., 2005; Adair, 2002; Watts and Bahill, 2000); one wonders if baseball fans are more likely to be scientists, or are scientists more likely to be baseball fans? A baseball's seam of over 200 stitches joins together two hourglass-shaped strips of leather. Although the ball is of a broadly similar size (9–9 1/4 inches circumference) and mass (5–5 1/4 ounces avoirdupois) as a cricket ball, its motions are in principle more complex since the stitching is not so simply arranged.

There are two principal pitches in baseball of aerodynamic interest, the curveball and the knuckleball. (Certain other pitches are named, such as the screwball, which is essentially a curveball with the spin axis reversed, and the slider, a fast pitch with the spin axis vertical. All are, in essence, just variants of a curveball—no different aerodynamic effects are invoked.)

Figure 2.9. Smoke trail photograph of a baseball spinning anticlockwise at 1000 rpm, flying at 60 feet per second to the right. The downwards deviation of the wake by 20 degrees or so implies significant Robins–Magnus lift. Photograph by Prof. F. N. Brown. Courtesy of University of Notre Dame Archives.

A less common pitch, perhaps, is the knuckleball, which is somewhat related to the swing bowl in cricket, in that the seam is used to trip turbulence asymmetrically. However, here the configuration of the seam does not permit a constant orientation by spinning. Instead, the pitcher attempts to throw with as little spin as possible. There is inevitably some rotation, which has the effect of causing the seam to be presented at a range of angles to the flow and thus cause a varying—and therefore hard for the batter to anticipate—side-force direction. It is estimated that a knuckleball may deviate by 27 cm from its initial trajectory before returning.

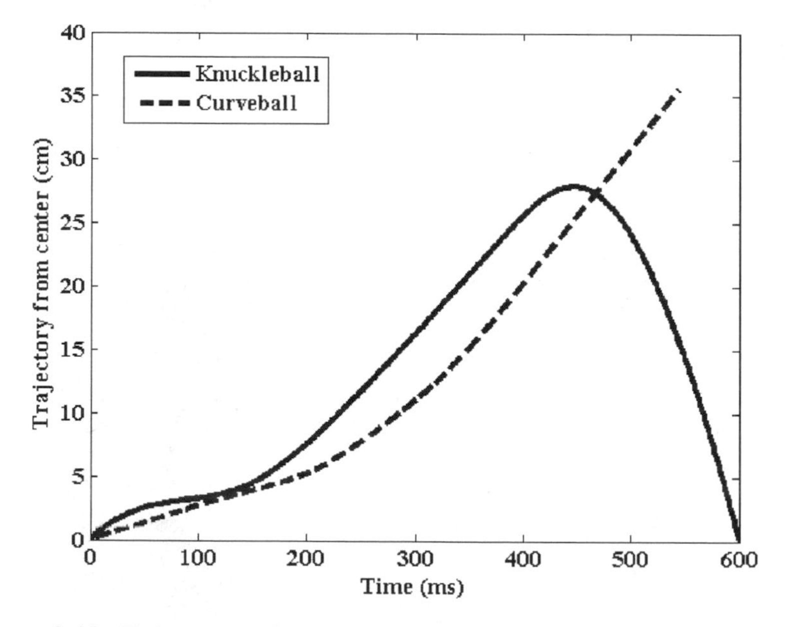

Figure 2.10. Trajectories of a curveball and knuckleball, showing the deviation from a centerline as a function of time—the ball traverses the 20 m from pitcher's mound to batter in around half a second.

A softball is of course somewhat larger than a baseball and will typically, as its name suggests, be thrown more slowly. Its size relative to the hand makes it much harder to impart spin to the ball. Thus, although slower pitches give longer times for side-forces to act, and low flight speeds give higher advance ratios for a given spin, in general spin effects on softballs are rather modest.

Table Tennis

Of all the sports considered here, the table-tennis ball is the lightest and smoothest. Aerodynamically it resembles a squash ball in size and smoothness, but has a mass/area ratio 10 times lower. It thus decelerates rapidly due to air. It is also more responsive to other aerodynamic forces such as the Robins–Magnus effect.

Saturn and the Squash Ball

Like a table-tennis ball, a squash ball is smooth. However, a squash ball has a much higher mass/area ratio, and thus its trajectory is less affected by aerodynamic forces, and is in fact nearly ballistic. The art of the game derives mostly from the kinematics of the bounce from the walls—use of multiple bounces rapidly eliminates the ball's kinetic energy. Topspin is used to cause a ball to bounce steeply downwards, making it hard to intercept and return in time, while backspin is used in the serve, to cause an upward bounce making for a steep descent at the back of the court.

Squash is notable in that the game's thermal component is very obvious—the coefficient of restitution of the ball is highly dependent upon the ball's temperature. Two effects are at work: the elasticity of the rubber, and the pressure of the air inside. How these affect the ball's bounce will be discussed in the next chapter.

A remarkable feature of a sliced (i.e., highly spun) ball in squash is that the ball becomes visibly distorted. In this respect, the squash ball is much like a planet; the spin introduces a centripetal acceleration which tends to draw the object into a flat disk, while a restoring force attempts to pull the object back into a spherical shape. In the case of the squash ball, the restoring force is the elastic nature of the ball material; in the case of a fluid planet, the restoring force is simply gravity.

The Earth has an oblateness of about 0.0034; while the equatorial radius of Earth is some 6379 km, the radius at the poles is only 6357 km. A planet with a given bulk density and rotation rate will have an oblateness which is constant with radius: the larger r, the larger the centripetal acceleration term ($\omega^2 r$), but also the larger the surface gravity, which depends on the mass (i.e., on the volume, and thus the cube of radius) divided by the square of distance.

Of the planets in the solar system, Saturn in fact has the largest oblateness (0.102), and it is visibly distorted. Its polar radius is 54,400 km, while the equatorial radius is 60,400 km. The planet's rotation period is only 10 hours and 40 minutes.

Figure 2.11. A color composite from images taken by the *Hubble Space Telescope* in 1995. The planet Saturn, with its rings edge-on, is perceptably flattened due to its spin. A few small moons are visible at the right, and Saturn's largest moon, Titan, is on the left, with its shadow cast onto Saturn. Image STScI/NASA/Erich Karkoschka.

Nonspherical Balls: (American) Football and Rugby

The stretched prolate shape of balls used in American football and in rugby introduce many complications. Most frustrating of these during play is the highly nondeterministic nature of the bounce—remarkable changes in direction can occur upon each contact with the ground. However, for the present chapter, the flight performance is more interesting.

The projected area of the ball is almost halved when the ball is end-on compared to when it is flying broadside. Thus, as is well known, the ball can be thrown further if thrown longwise. The ball is not statically stable (at least not appreciably so) and thus a spin must be imparted to the ball in order to keep its long axis pointed in the direction of motion. Thus most good passes are "spiral" in nature. Video and flight data recorder measurements suggest 600 rpm is a typical spin rate for a thrown football.

Because of the tendency for aerodynamic moments to precess the ball, and the intrinsic instability of rotation around the axis of minimum moment of inertia, the ball tends to begin nutating, if not tumbling, towards the end of the throw. Some of the ball's angular momentum is shed into the wake, which would ultimately cause the spin to decrease.

Figure 2.12. Simulation of the velocity around a spinning football, flying left to right. Courtesy of Fluent, Inc.

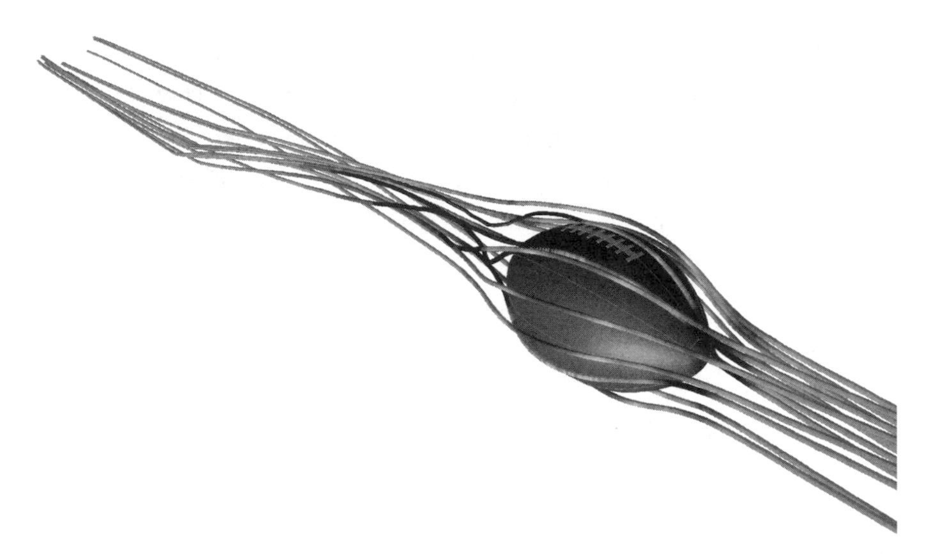

Figure 2.13. Streamlines around a spinning football flying towards lower right—note the twisting, indicating that angular momentum is being shed into the wake, slowing down the spin. Image courtesy of Fluent, Inc.

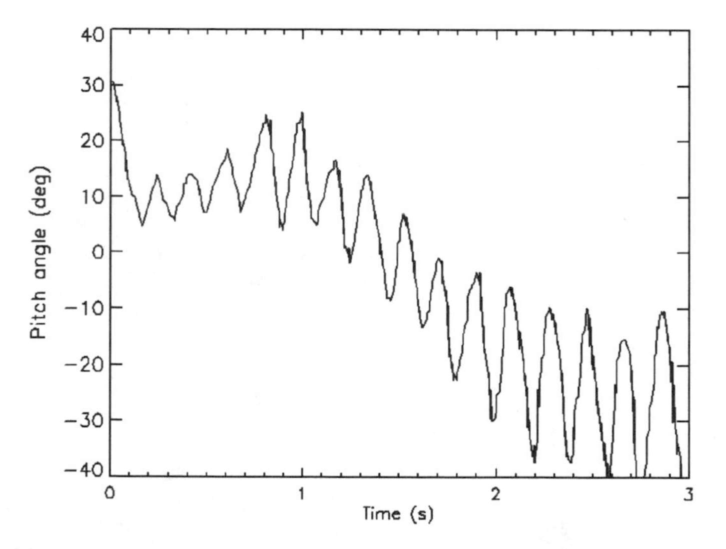

Figure 2.14. The pitch angle (angle between the long axis of the football and the horizontal), from video measurements by Rae (2003). The mean angle (i.e., the spin axis) turns with time due to yaw moments, while the envelope of the oscillating curve increases due in part to loss of energy while largely conserving angular momentum.

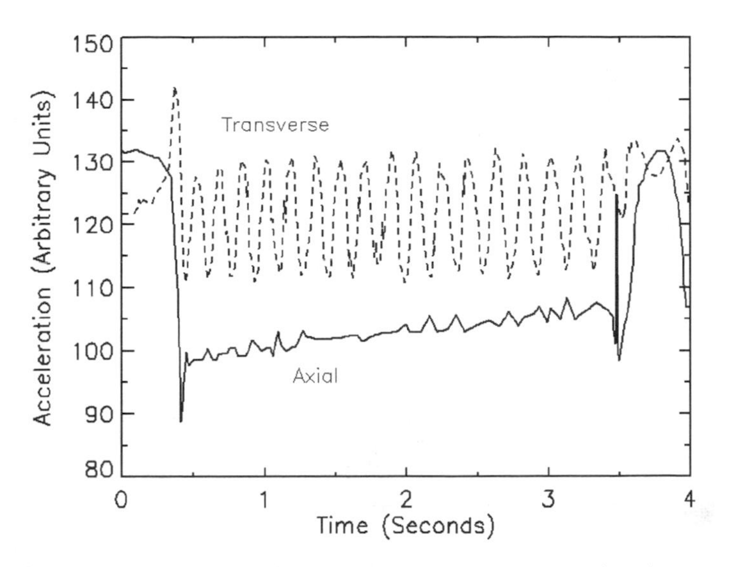

Figure 2.15. Measurements of the axial and transverse accelerations recorded by an accelerometer and data-logging package embedded in a foam-filled American football by Nowak et al. The axial acceleration is somewhat constant, declining as the ball slows in flight (and perhaps also because the axis begins to cone around the direction of flight). The transverse acceleration is modulated by the spin (~ 6 revolutions per second, slowing towards the end of flight). Compare these data with those from a Frisbee in chapter 8.

The moments of inertia of a football are such that this tumbling, as measured by in-flight accelerometer measurements (Nowak et al., 2003), has a period 1.8 times the initial spin period. The tumbling (or rather, nutation) period is longer than the spin period for long objects; for flat discs the nutation period is close to half of the spin period.

While it is possible to introduce spin about the long axis by good throwing technique, such a spin cannot be easily administered with the foot. The momentum that can be transferred to the ball by a kick is rather more than is possible from a throw, but the flight must then take place without the benefit of a minimum-drag orientation. A very carefully placed kick might be able to avoid significant tumble on the ball, but then the trajectory would be susceptible to the body-lift generated by the ball which might be at some modest angle of attack throughout the flight. Accordingly, normal practice is to kick the ball such that it has a consistent end-over-end tumble, "rolling along" its flight path. This has the advantage of presenting a cross-section to the flow that is at least sometimes close to the minimum, resulting in a lower average drag than would be likely without tumble. However, it may be that the achievement of a successful kick (considering the contact kinematics of the ball and foot) is the driving factor in a kick, rather than the subsequent flight performance.

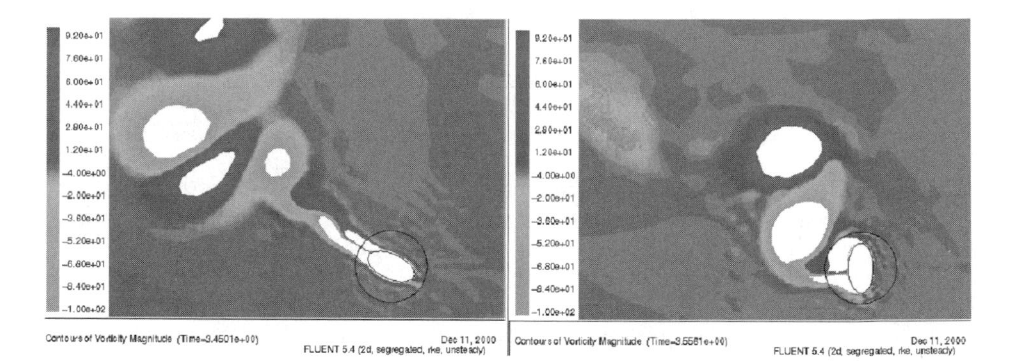

Figure 2.16. Images of the vorticity shed by a tumbling rugby ball in flight. The wake is quite narrow (left) when the ball is nearly aligned with its direction of flight (to lower right), but large vortices are shed (right) as the ball rotates to become temporarily sideways-on. Images courtesy of Fluent, Inc.

Studies (e.g., Watts and Moore, 2003) suggest the axial force is around 0.14 at zero incidence (at this angle, the axial force coefficient equals the drag coefficient), with a cosine variation down to zero at 90 degrees. Rae and Streit (2002) note the difficulty in making force measurements on a ball which is rapidly spinning—the ball must be mounted on a motorized support or "sting", and any slight imbalance in the mounting of the ball (which is somewhat massive compared to many wind tunnel models) severely degrades the force data. Aerodynamic forces have to be inferred from the difference between strain gauge measurements made with the airflow on in the tunnel, and with airflow off. This difference measurement relies on the mounting being undistorted by the airflow, a challenging assumption.

Rae and Streit measure a side-force coefficient as a function of angle of attack; this varied up to around 0.45 at an angle of attack of 50 degrees and a spin rate of 600 rpm. At 100 rpm and the same angle of attack, the coefficient was around 0.2, suggestive of a nonlinear dependence.

The pitch moment coefficient on the ball varies linearly with angle of attack and is essentially invariant with spin: both wind tunnel measurements and video observations suggest it has a slope of around 0.005 per degree, flattening out to around 0.18 above 30 degrees. The normal force coefficient varies linearly with angle of attack with a slope of 0.013 per degree.

Conclusions

A variety of techniques have been used to explore the dynamic and aerodynamic properties of sports balls to an ever-improving degree. Techniques have ranged from judgements by eye, to wind-tunnel tests and computational fluid dynamics, to measurements with tape—paper and video!

As these investigations have led to higher-performing balls, sports have been significantly affected, to the degree that modifications to decrease performance, such as slower tennis balls, are being contemplated by the relevant regulatory bodies.

Many balls operate at or near a critical Reynolds number, such that drag coefficient can be a strong function of flight speed. Substantial side- or lift-forces are developed by many balls via seam-triggered boundary layer transition and thus delayed separation; in other cases Robins–Magnus lift plays the dominant role.

References

Adair, R., *The Physics of Baseball*, 3rd Ed. Harpercollins, 2002 (Note: this book is written for baseball fans, not physicists—it uses English units, and several aspects, notably the discussion of the Magnus force, could be better.)

Bahill, A. T., D. G. Baldwin, and J. Venkateswaran, Predicting a baseball's path, *American Scientist*, 93(3), 218–225, May–June 2005.

Barkla, H., and A. Auchterlonie, The Magnus or Robins effect on rotating spheres, *Journal of Fluid Mechanics*, 47, 437–447, 1971.

Bearman, P. W., and J. K. Harvey, Golf ball aerodynamics, *Aeronauts Quarterly*, 27, 112, 1976.

Carre, M. J., T. Asaum, T. Akatsuka, and S. J. Haake, The curve kick of a football II: Flight through the air, *Sports Engineering* 5, 193–200, 2002.

Cooke, Alison J., An overview of tennis ball aerodynamics, *Sports Engineering*, 3, 123–129, 2000.

Mehta, R. D., Aerodynamics of sports balls, *Annual Reviews of Fluid Mechanics*, 17, 151–189, 1985.

Nowak, Chris J., Venkat Krovi, William J. Rae, Flight data recorder for an American football, *Proceedings of the 5th International Conference on the Engineering of Sport*, Davis, California, September 13–16, 2004.

Rae, W. J., and R. J. Streit, Wind tunnel measurements of the aerodynamic loads on an American football, Sports Engineering, 5, 165–172, 2002.

Rae, W. J., Flight dynamics of an American football in a forward pass, *Sports Engineering*, 6, 149–164, 2003.

Lord Rayleigh, On the irregular flight of the tennis ball, *Mathematical Messenger* 7, 14–16, 1878.

Stepanek, Antonin, The aerodynamics of tennis balls: The topspin lob, *American Journal of Physics* 56, 138–142, 1988.

Tait, P. G., Some points in the physics of golf, *Nature* 42, 420–423, 1890. Continued in 44, 497–498, 1891 and 48, 202–204, 1893.

Thomson, J. J., The dynamics of a golf ball, *Nature* 85, 251–257, 1910 Volume 3, Issue 2, 123, May 2000.

Watts, R. G., and G. Moore, The drag force on an American football, *Am. J. Phys*, 71, 791–793, 2003.

Watts, R. G., and A. T. Bahill, *Keep Your Eye on the Ball*, Freeman, New York, 2000. (A rather more physically rigorous book than Adair.)

http://www.hq.nasa.gov/alsj/a14/a14.html

http://news.bbc.co.uk/sportacademy/hi/sa/tennis/features/newsid_2997000/2997504.stm

3

Bouncing Balls, Airbags, and Tumbleweeds

Flights need not consist of uninterrupted transit through the air. An important aspect of many ball trajectories is their interaction with the ground: bouncing. Furthermore, the effects of spin on ball trajectories only come about because spin can be induced. In many cases the spin is induced by the bouncing or striking geometry with a bat or racquet. The physics of bouncing and rolling balls also becomes relevant to an experimental type of planetary exploration vehicle, the Tumbleweed rover.

When two objects collide, their kinetic energy is converted into deformation of one or both objects, with the deformation depending on the compliance (i.e., "softness") of each. If one object is more compliant than the other, then it will deform more; in many cases one object is so much more compliant that only its deformation need be considered. The other object, often a wall, may be considered completely rigid.

The deformation may or may not be elastic (in fact, the second law of thermodynamics more or less requires that no macroscopic impact can be completely elastic). In a purely elastic impact, the object wants to spring back to its original shape, and the energy stored in its deformation is converted back into kinetic energy with 100% efficiency. If, however, some of the energy is converted into heat and lost, or expended in irreversible ("plastic") deformation of the material, then kinetic energy is dissipated and the objects will bounce apart with a lower speed than that with which they collided.

Sometimes this loss is characterized by a number called the *coefficient of restitution* (COR, or e), which is the ratio of the separation speed to the collision speed. Since kinetic energy is proportional to the square of speed, the efficiency (i.e., the fraction of the conserved kinetic energy) is equal to the square of this coefficient.

The Bouncing Process

But what actually happens when a ball bounces on a hard surface? Obviously, the ball must flatten somewhat, and then regain its shape. Its direction reverses, so during the contact period the ball exerts a force on the surface and the surface exerts an equal and opposite force on the ball.

There are two factors at work to make the ball exert a force: the elastic properties of the material from which it is made, and the air pressure (if any) inside it. Clearly, the relative importance of these forces depends on the ball.

One set of objects are pure solids—marbles, billiard (pool and snooker) balls, and solid rubber balls such as the "Super Ball" (coincidentally invented by Ed Headrick, who as we shall see in a later chapter also developed the modern Frisbee). Cannonballs also come under this category.

Another category are more or less thin-walled spheres, where the elastic properties of the ball material are the dominant factor in the bounce. Table-tennis (Ping-Pong) balls are a prime example; tennis balls and cold squash balls also come broadly under this category. Soap bubbles are another exotic example.

Finally there are pressurized spheres: balloons, beachballs, and the airbags used in spacecraft: landing are examples, as is a warm squash ball. Here the wall material is so thin, or so compliant, that it exerts a negligible force. However, the pressure excess of the inside of the ball above its environment (which is what, after all, gives the object its spherical form) will act over the contact area to develop a force against the surface.

A variety of techniques have been used to investigate the bouncing process; because balls are easy to procure and are familiar, many experimental exercises in high school mechanics are illustrated by ball experiments. High-speed video data has been used to show the elastic deformations of the ball, although in some cases a regular camcorder at 25 frames per second is adequate to measure the bouncing trajectory of a ball and infer the collision dynamics (e.g., Cross, 2002). Rather lower-tech, but nonetheless ingenious, experiments have used inked balls to mark a surface, and thus show the contact area reached during the impact (Bridge, 1999). Balls with conductive (foil or paint) plates have been bounced on plates with exposed copper tracks such that the tracks are bridged, making electrical contact when the ball is in contact with the plate.

Among the most efficient methods has been the use of load cells or piezoelectric sensors, mounted under the surface from which bounce will take place (e.g., Carre et al., 2004; Cross, 1999). These sensors allow the measurement of the force vs. time, even during the couple of milliseconds typical of an impact.

A force–deformation curve can be drawn using data from a piezoelectric sensor or other load cell onto which the ball is dropped. Such curves are also readily generated from presses, although here the process is very slow and the characteristics are somewhat different.

The curve forms a loop, with the compression during impact going one way, and release and rebound going back to the origin. More often than not, the compression and release curves are not coincident; the latter leg is lower. This means there is a gap between the two, which implies that energy is lost—hysteresis. This energy loss is manifested in the coefficient of restitution. Bodies with high coefficients of restition show nearly coincident force–deformation curves; little energy is lost. The rubber superball is a good example.

At the other extreme are balls which do not bounce at all. Plasticine is one possibility—it deforms plastically, with no bounce. On the force–deformation plot, it requires a force to be compressed, but on release it does not spring back.

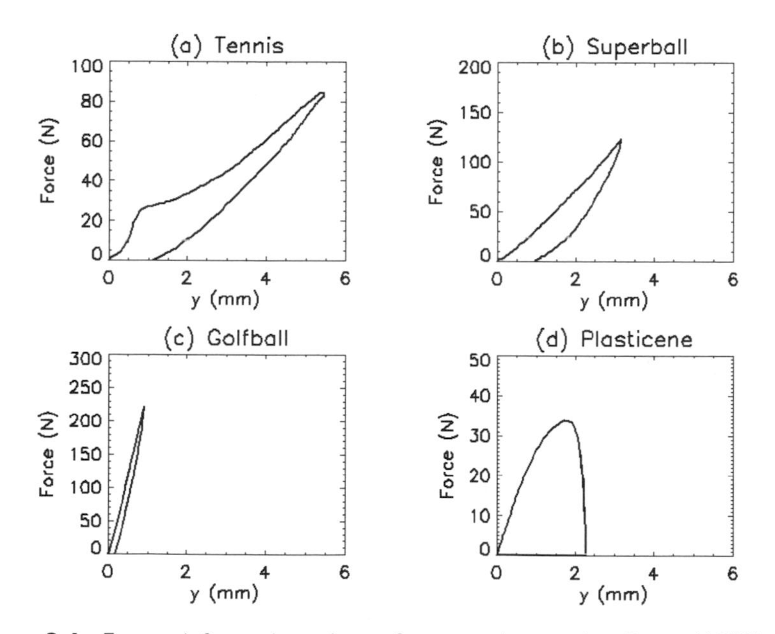

Figure 3.1. Force–deformation plots, after experiments by Cross (1999) with impact speeds of 1.5 to 3 m/s. Note the different scales; the golfball is much harder than the others. Note also the area enclosed by the curve; this area represents the energy lost by the ball in the collision. The inelastic plasticene ball is the worst offender in this regard. Note also the change in slope in the tennis ball curve; this "softening" is due to the buckling of the sphere after the initial compression.

Bounce, Spin, and Skid

The bounce process is one of energy conversion. The ball may have translational kinetic energy due to velocity components parallel and normal to the reflecting surface (V_x and V_y, respectively), and it may have rotational kinetic energy. At impact some of this is converted into elastic potential energy by the deformation of the ball.

In a purely vertical impact, translational kinetic energy is completely converted into elastic deformation, and some fraction (e^2) of this elastic potential energy is restored. More generally, however, there can be an exchange between all three energies, such that after the bounce is completed (and elastic energy is zero) some energy has been exchanged between the rotational and translational kinetic energies. Although one can at first describe the surface in terms of two coefficients of restitution e_x and e_y, a complete description (e.g., Cross, 1981) of the energy exchange also requires knowledge of the coefficient of friction of the surface, the radius, and the moment of inertia of the ball. In general the spin–impact coupling effects are strongest when the ball is soft, such as a warm squash ball. The obvious counterexample is the billiards ball, where little deformation occurs.

An example of the role of the coefficient of restitution"s importance is in tennis, where there is a marked difference in style of play on grass (as at Wimbledon) and clay courts (as in the French Open).

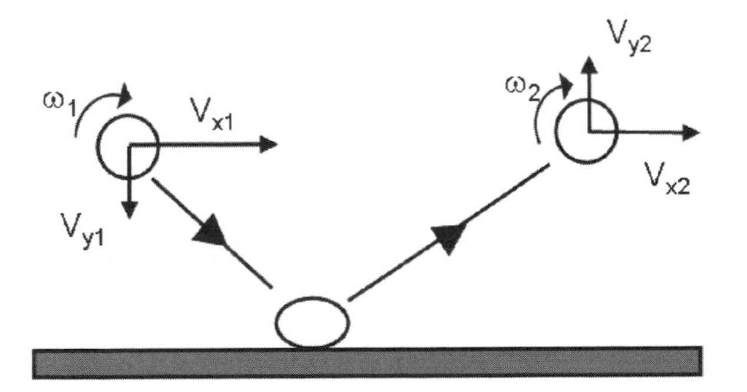

Figure 3.2. Collision geometry. The conventional (vertical) coefficient of restitution e_y is simply V_{y2}/V_{y1}. The horizontal coefficient of restitution is more complex, $e_x = V_{x2} - R\omega_2/(V_{x1} - R\omega_1)$.

Table Tennis

Table tennis balls have very thin walls compared with their diameter. Their ability to bounce without becoming completely flat requires a rather low compliance. In order to achieve this with a thin wall requires that the material itself be rather stiff, to have a high Young's Modulus. Thus, table tennis balls are made of nitrocellulose (the same material as billiard balls) rather than rubber.

High-speed photography (Hubbard and Stronge, 2001) taken through a glass plate onto which a ball was bounced shows that a table tennis ball not only flattens itself against a surface, but the impacting side of the ball pops inwards. Once the deformation becomes sufficiently high, a "popped" shape becomes energetically favorable compared with a flat shape. As the ball bounces away from the surface, it "pops" back. The fact that the ball pops inwards in this way means a significant drop in volume, and thus a spike in the internal pressure.

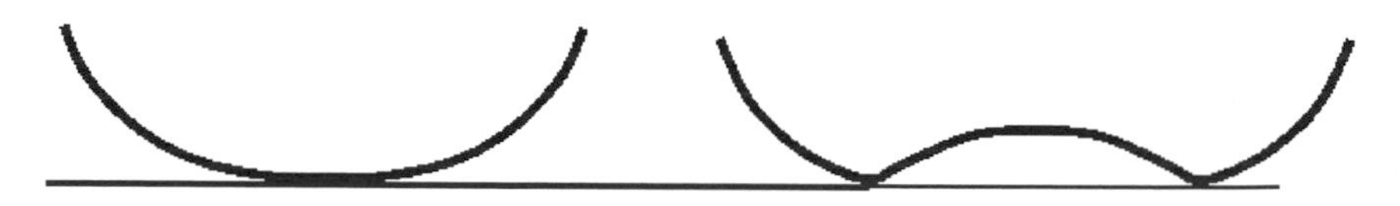

Figure 3.3. Schematic of the impact of a ball with elastic walls on a hard surface. After some initial flattening, the compression around the contact circle becomes large enough that a flat shape is energetically unfavorable, and the ball pops inwards. This leads to a drop in the slope of the force–deformation curve (i.e., the stiffness drops).

Tennis Balls

Tennis balls are specified by their size and weight, and by their ability to bounce between 53 and 58 inches when dropped on a concrete slab from 100 inches up. This implies a coefficient of restitution e in the range $0.728 < e < 0.762$.

Various designs are used, some with internal pressure (these are shipped in sealed, pressurized cans to minimize leakage), some without. The unpressurized variants have thicker walls to provide extra stiffness.

High-speed video of tennis balls shows that wave oscillations travel across their surfaces during collisions, excited perhaps by the inward popping of the skin at impact. These waves, as well as creating instantaneous deformations of the skin of up to 1 cm, are responsible for the sound of the bounce; pressurized balls damp the characteristic 1.2 kHz oscillation more quickly, and the ball sounds "duller."

The inversion of the ball surface has been neatly demonstrated by Cross (1999), who used a small set of piezoelectric sensors to measure the instantaneous force at several locations on the surface being impacted. The center sensor sees a dip in the middle of the impact, suggesting that the surface has popped inwards and no longer pushes on the wall. Away from the center, however, the ball surface is still in contact with the wall (see Figure 3.4)—the situation is as in Figure 3.3.

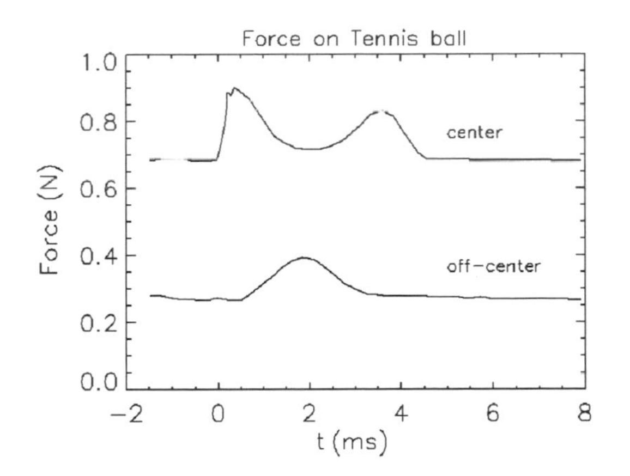

Figure 3.4. Force history of the collision of a tennis ball with an array of piezoelectric sensors, after Cross (1999).

Although the ball construction is not tightly regulated in tennis, the rackets are given scrutiny. The racket significantly affects the dynamics of play. As discussed by Brody (1979), the coefficient of restitution of a ball dropped on the strings of a tennis racket is higher than if it is dropped on a hard surface—the strings are less compliant, but more elastic, than the ball itself. The "sweet spot" (Brody, 1981) may or may not be the point on the strings where the COR is maximized; another definition is the node of the structural oscillations of the racket—the place where the vibrations are a minimum. Much of the effort in racket design is aimed at reducing these vibrations.

Perhaps even more important than the influence of the racket on the bounce normal to the string plane is the racket's influence on spin (e.g., Goodwill and Haake, 2004). Specifically, the strings are designed to stretch, but are not supposed to move along one another (i.e., within the string plane). Cross strings are alternately woven above and below the main strings to prevent their relative movement. Such movement, seen in so-called "spaghetti strings" in 1970s rackets, allows too much spin to be put on the ball, and the International Tennis Federation accordingly bans these strings.

Squash Balls

It has been observed in experiments (Chapman and Zuyderoff, 1986; Bridge, 1998) that the contact time of a squash ball increases with temperature, or equivalently that the ball gets softer. This implies that the softening of the rubber (which increases the contact time) is a stronger effect than the increase of the air pressure inside (which would decrease the contact time).

It is well known to squash players that the ball bounces better when warm, and some effort is usually expended at the beginning of a game just to warm it up. The proportion of the impact force due to air pressure (which is elastic) increases at high temperature, since the (inelastic) rubber wall becomes unable to provide the necessary force when warm. Thus the net effect is a softer ball (in terms of requiring more deformation for a given force) but a more elastic one.

The coefficient of restitution of a squash ball decreases substantially with the speed at which it hits the wall—at low velocity (~5 m/s) the standard "Yellow Dot" ball has $e \sim 0.45$ at 25 °C, increasing to about 0.6 at 41 °C. However, for really hard shots (~60 m/s) e is only around 0.2 for all temperatures. The slightly less elastic Double Yellow Dot balls—sold for use in warm conditions, or by advanced players—have low-speed restitution about 0.05 smaller than the Yellow Dot.

Superballs

These rubber spheres are of interest as intriguing toys with a high stiffness and coefficient of restitution. The coefficient of restitution—which Cross (2001) determined to be around 0.86 for low speed impacts—decreases for old balls, which tend to have cracks as the rubber ages and becomes more brittle. The coefficient decreases for faster impacts.

The high coefficient of friction—Cross found 0.52 for a superball on wood, compared with only 0.15 for a tennis ball on the same surface—leads to the rather curious dynamic property that a superball often bounces back and forth on a surface, rather than bouncing forwards. Tokieda reports that usually two or three back-and-forth bounces can be generated in practice.

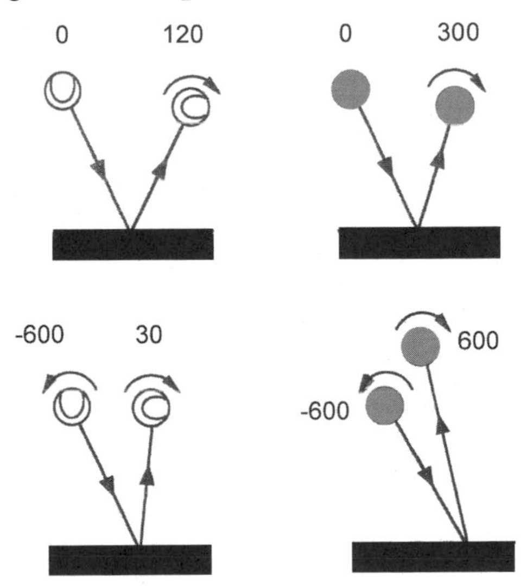

Figure 3.5. Comparison of the bounce of a tennis ball (left) with a rubber superball (right). Rotation speeds are indicated by arrows, with speed in rpm indicated with no-spin (top) and backspin (bottom) cases shown. The superball's contact friction with the wall is enough that the backspin can reverse the translational direction of the bounce.

Another property of the high friction combined with the high coefficient of restitution is that a ball bounced on the floor under a table can be "reflected" back (Garwin, 1969) to the thrower, even though there is no vertical surface to reverse the ball"s horizontal motion.

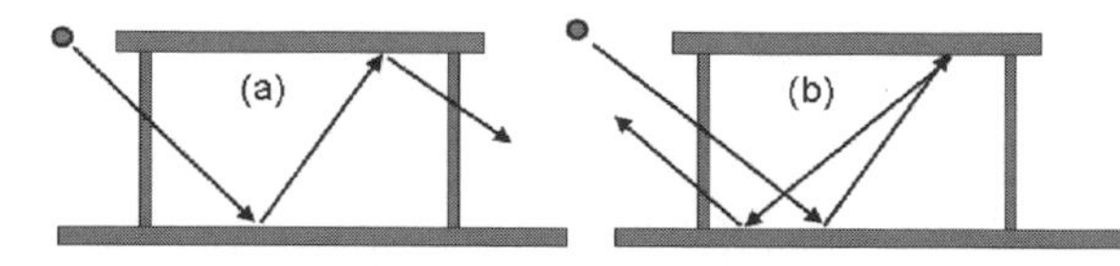

Figure 3.6. A ball thrown to bounce on the underside of a table will, for low friction (a), keep moving in the same horizontal direction. A high friction ball (b) can be "reflected" due to the spin induced at the first bounce.

Levitating Golf Balls

Perhaps one of the most violent ball collisions is that between a golf club and ball. Golf is such big business that this collision has been quite well studied by modern methods, e.g., Iwatsubo et al. (2000) and Roberts et al. (2001).

However, one aspect of rolling ball behavior that is so perverse that one is inclined to dismiss it as just bad luck is that of a ball rolling inside a vertical cylinder. This seemingly abstract scenario is what occurs when a golfer sinks a ball into the hole. One might expect the ball to cooperatively describe a spiral descending motion before hitting the bottom, at which point it might bounce a little, but would nonetheless remain in the hole. And yet, the ball often appears to defy this expectation by flying back out of the hole without hitting the bottom.

This behavior has recently been systematically documented by Tadashi Tokieda, Marco Gualtieri, and colleagues (Gualtieri et al., 2005). They elegantly show how a ball rolling in a cylinder will undergo vertical oscillations, which they attribute to a Coriolis torque. I confess I don"t fully understand this myself, but they derive (for a uniform sphere—the result depends on the moment of inertia of the ball) that the periods of vertical and horizontal oscillation should be in the ratio $(7/2)^{0.5} \sim 1.87$. They conduct experiments with a plexiglass cylinder and a computer mouse ball (a suitably dense ball, its rubber coating giving good strong friction with the cylinder surface), videotaping the trajectory of the ball. In some instances the ball oscillates up and down as it spins around the inside of the cylinder, while in other cases it escapes from the cylinder altogether (i.e., the amplitude of the oscillation is larger than the depth at which the ball was launched into circumferential motion). For the mouse ball moment of inertia, the ratio of oscillation periods is almost exactly 2.

The question then arises of how to prevent a golf ball from jumping out of the hole. Making the ball of smaller diameter and the hole larger would work, but this seems to be against the spirit of the game. Reducing the friction between the ball and the hole would help, as would increasing the radius of gyration (i.e., the moment of inertia per mass) of the ball. However, the only practical way is just to hit the ball slowly enough.

The way I think of it is that the ball is rolling at first horizontally inside the cylinder, so its spin axis is vertical. As gravity pulls the ball down, the rolling path descends, and thus the friction acting at the contact point begins to have a small upwards as well as horizontal-backwards component. This upwards friction generates a small torque which precesses the spin axis back such that the ball rolls slightly upwards.

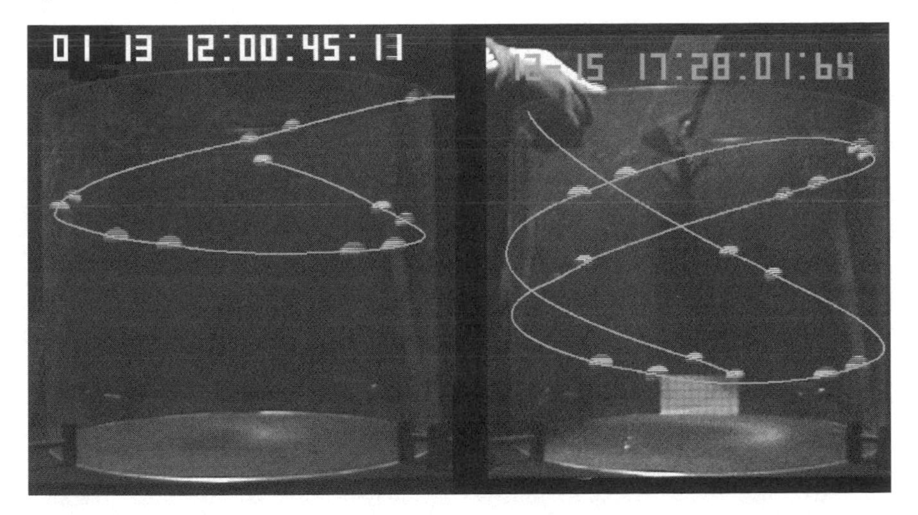

Figure 3.7. Frames from a video record showing the trajectory of a ball rolled on the inside of a vertical plexiglass cylinder. The slightly striped appearance of the ball is due to the interlaced video signal. In the image on left, the ball makes just over one circuit, during which it reverses its vertical motion. On the right the ball rolls down, up, and back down again, oscillating in a vertical plane roughly once in two horizontal revolutions. Image courtesy of Erwan Reffet, C. Guthmann and colleagues—details in Gualtieri et al., 2005.

Pathfinder Airbag

Bouncing and rolling turns out to be an occasional mode of locomotion in planetary exploration—indeed, all of the last three successful Mars landers have bounced and rolled to a halt at the end of their hundred-million-mile trip.

Some of the earliest ideas for landing spacecraft on the moon or other planets relied on the use of some sort of shock-absorbing material like a crushable honeycomb or even balsa wood. The basic idea is that for a given impact speed, the longer the distance over which one decelerates, the lower the deceleration. Thus a thick layer of soft material would allow the equipment to be decelerated by impact only rather slowly. Since the orientation of the lander could not be easily controlled, the impact attenuation material had to cover all directions, giving a spherical geometry. In fact, the first U.S. Ranger satellites to the moon carried small balsa wood capsules which were to be released for safe arrival just before the main spacecraft made their destructive impacts. However, the early Rangers were plagued with faults and none of the capsules were delivered correctly.

The concept of a simple shock-absorbing landing rather than using legs with retro-rockets and all the paraphernalia required to control them was invoked for Mars in the early 1990s, when a seismology and meteorology network mission was planned, named Mesur. This would have deployed many small landers at different points on Mars, and so would need a much cheaper landing system since over 10 would be built. The idea emerged of using an Airbag—a deceleration system using a bag pressurized very quickly with gas from a chemical gas generator—ust like those used in many cars nowadays. In the end, Mesur faded away, being seen as too expensive, but a proof-of-concept flight named "Pathfinder" (originally Mesur Pathfinder) was flown to show that the idea could work.

This spin-stabilized spacecraft entered the Martian atmosphere and descended by parachute. Shortly before impact, a set of airbags were inflated and a small rocket fired to take away the parachute, leaving the airbag-encased lander to fall to the surface and bounce and roll.

The airbags were designed to limit the deceleration on the 230 kg lander to 50 Earth *g*: given the initially specified impact speed of 35 m/s, this requires a deceleration stroke of 1.25 m. In addition, the mission was designed to tolerate rocks of up to 0.5 m, so the airbags had to be at least 1.75 m thick. The airbags were made of a Kevlar fabric with a urethane coating; vents between the bags allowed gas to be transferred between the bottom bag and the upper ones during impact (Waye et al., 1995).

Figure 3.8. The *Mars Pathfinder* lander in a ground test, encased in its cluster of airbags. NASA image GPN-2000-000484.

This vehicle, the first lander to use airbags for impact attenuation, hit the Martian surface (slowed by a parachute and retro-rockets fired just before release above the impact site) with a vertical velocity of 14 m/s and an estimated horizontal velocity of 6 m/s; the loads on the first bounce were 18.7*g*. It is estimated (Spencer et al., 1998) the lander traveled about 1 km in the subsequent 2 minutes of bouncing—several initial bounces were captured by the accelerometers. Following *Pathfinder*'s success, the airbag technique was subsequently used on the two Mars Exploration Rovers *Spirit* and *Opportunity*.

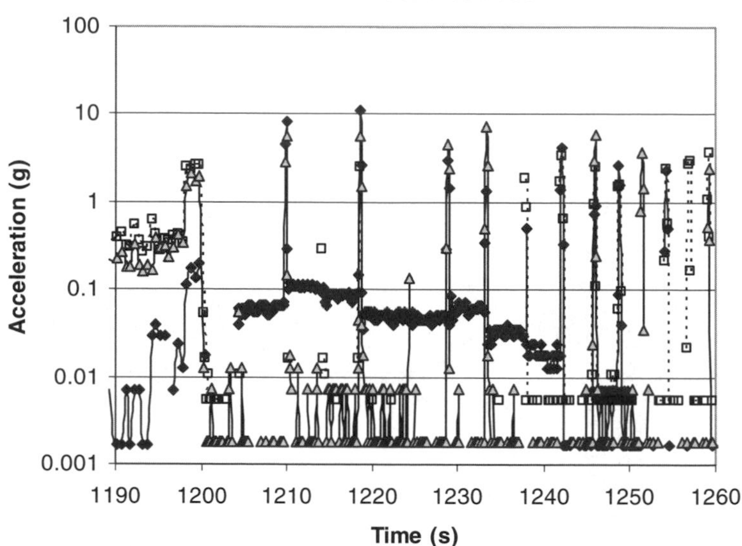

Figure 3.9. 3-axis accelerometer record (note logarithmic scale) of the *Pathfinder* impact and subsequent bounces on the surface of Mars. The acceleration spikes become progressively weaker and closer together as the bounce height decreases—energy is steadily being lost due to drag in the atmosphere and the inelastic bounces. Note also the steady-state ~0.1 to 0.01 g signal in one axis—this is presumably due to a centripetal acceleration, with the accelerometer offset from the center of mass. Bouncing was still continuing when the record timed out at 1260 s, a minute after first contact.

Tumbleweed: The Russian Thistle

Bouncing and rolling turns out to be an occasional mode of locomotion for plants on Earth. The tumbleweed, also known as the Russian thistle or saltwort, is familiar in films of the Wild West. It is more formally known as the prickly Russian thistle—*Salsola kali ruthenica*—and it exploits wind to accomplish dispersal, not of individual seeds, but of the entire plant. The plant was introduced into the Americas in the 1800s by immigrants from south and western Russia.

Figure 3.10. A tumbleweed perched on the rim of Meteor Crater, Arizona. Notice how the branches curve away from a common point, leading to an offset between center of mass and center of area. Photo by Eric Palmer, used with permission.

This annual plant is a rounded bushy cluster of branches, growing from 0.3 to 1 m in height and from 0.3 to 1.7 m in diameter. After a growth phase with long leaves through spring, these leaves fall off and are replaced with shorter, broader leaves. The plant has small whitish (wind-pollinated) flowers in late summer and autumn. These flowers are replaced with winged, 2 mm grey or brownish seeds which are retained by the plant until it dies. When this happens, the plant stem separates from the root via a special set of cells and the dry bundle of seeded branches is blown by the wind, scattering the seeds far and wide. The seeds, of which many tens of thousands may be released by a single plant, may be scattered further after being dropped from the tumbling plant.

It is worth noting that the center of mass of a tumbleweed is generally offset from the geometric center. To what extent this is an inevitable consequence of the dendritic architecture of a plant, in that the branches must converge towards an apex which is linked to the root system, is unclear. It may be that there are dispersal performance advantages in such a departure from spherical symmetry; tumbling/bouncing rather than rolling may enhance the shedding of seeds from the plant.

Tumbleweed Rovers

A Tumbleweed rover is a quasi-spherical vehicle intended to traverse a planetary surface (nominally Mars) with a rolling and/our bouncing motion driven by the wind. The name derives from the similar motion of the tumbleweed plant.

In the technological application, wind-driven rolling offers the potential for very long traverses across a planetary surface, albeit with little or no control over direction. It should be recalled that even the highly successful Mars Exploration Rovers that arrived in early 2004 each only traversed less than 5 km in an entire year of operation.

Figure 3.11. Engineer Alberto Behar with a Tumbleweed rover prototype in the Mohave desert. Note the diamond pattern of a web of cords intended to provide both abrasion resistance and traction, and a preferred roll axis via larger moment of inertia. Image used courtesy of NASA/JPL/Behar.

Two origins have been cited for the concept. A group at NASA Langley (Antol, 2003) reports being inspired by the bouncing motion of the *Mars Pathfinder* lander during its airbag landing in 1997. The idea naturally followed of seeing how far it might go if the airbags were never deflated.

The second origin is attributed to tests near another NASA center, the Jet Propulsion Laboratory in Pasadena, California. Here, a group was exploring mobility enhancement by using very large inflatable wheels; the size of obstacle that can be cleared by a vehicle typically relates to the wheel diameter. Experiments were being conducted with a three-wheel rover on a desert playa, a hard and very flat expanse. The rover's wheels were approximately spherical, some 1.5 m in diameter. According to the story (Jones, 2001; Matthews, 2003), one of these wheels came off or had been removed for adjustment when it started to blow in the wind, and indeed required speedy chase in a pickup truck to be recovered. This experience then prompted the idea of doing away with the rover frame and motors altogether.

As with many ideas, related concepts have surfaced earlier. One 1980s architecture was the University of Arizona "Mars Ball" wherein an inflatable pseudospherical vehicle might move across the Martian surface by sequentially deflating and inflating segments, and the idea of such a vehicle being blown around in the wind was advocated by planetary ballooning pioneer Jacques Blamont.

Tumbleweed designs discussed to date have tended to assume an equipment module suspended in the center of the vehicle, most usually an inflated structure although some deployable rigid structural designs have been considered (see disussim at the end of this chapter.). I myself have advocated using equipment integrated into the spherical structure itself, since more and more electronics can be made into thin or even flexible packages.

Tumbleweed rovers scale up strongly—larger balls are able to traverse rougher terrain without getting stuck, and have a larger drag area and thus typically begin moving at lower threshold windspeeds. Note that the analyses applied to Tumbleweed motion to date are rather inad-

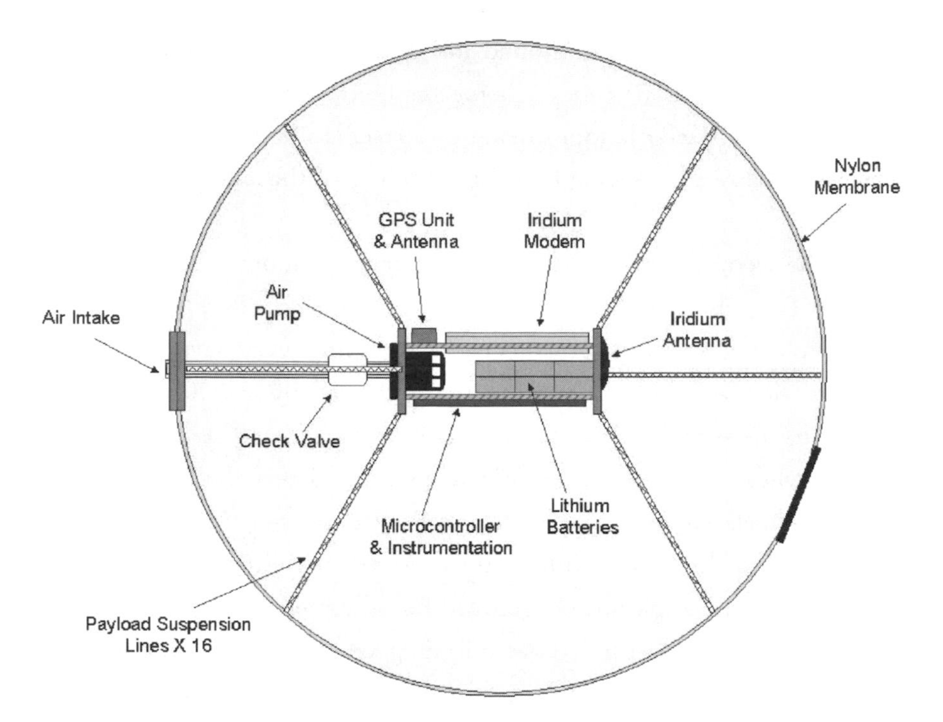

Figure 3.12. Cross-section of a testbed Tumbleweed, with an equipment package suspended inside a ball by elastic cords. The package is powered by batteries, and uses a pump to maintain the inflation pressure of the sphere. Image from Alberto Behar/JPL, used with permission.

equate, considering essentially only a drag force. In fact, in common with sediment transport by air and water on Earth, a Tumbleweed's motion has a significant lift component; the ball is not immersed in a uniform flowfield but is in the boundary layer with a typically logarithmic wind profile, slowest at the ground and increasing with height. Figure 3.13 shows such a profile measured on Mars by *Pathfinder*. An additional complication for the real tumbleweed and some of the non-inflatable designs is that, like many parachutes, the tumbleweed has a structure that is porous to the flow, and thus flow through as well as around the structure must be considered (Figure 3.14).

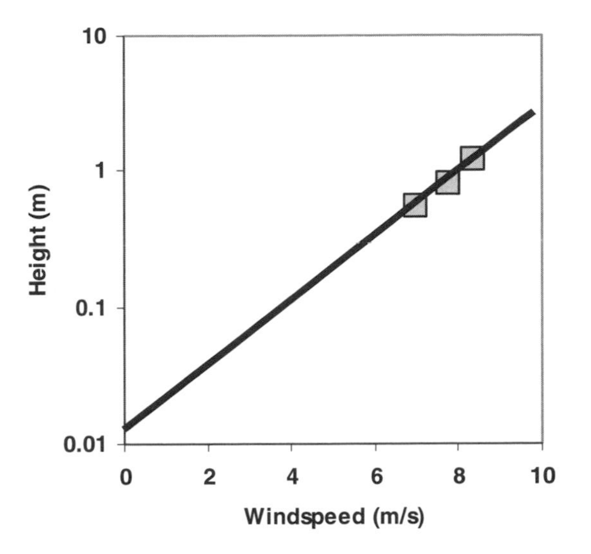

Figure 3.13. Wind profile near the Martian surface measured on Sol 52 by the windsock experiment on the *Mars Pathfinder* lander (data from Sullivan et al., 2000). The windsocks between about 0.5 and 1.2 m above the ground measured a variation with height that follows a classical logarithmic profile (a straight line on this plot). The intercept at zero speed implies an "aerodynamic roughness" of about 0.02 m.

The flow over the top of the rover therefore creates suction—a lift force. This can be enough to pluck the ball off the surface outright, but more usually simply reduces the rolling friction. It may be that roughness elements on the ball's surface also allow the boundary layer flow to spin the ball up, imparting a torque as well as lift and drag.

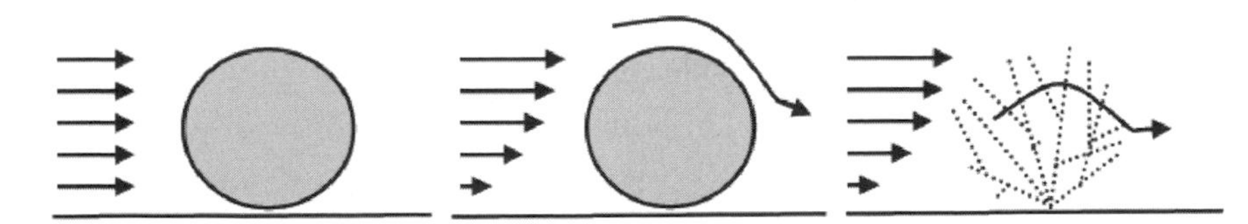

Figure 3.14. Flow around a Tumbleweed rover. On the left is the scenario usually considered analytically, of a uniform drag on a sphere. Center is the more probable situation, with the rover immersed in a boundary layer with a logarithmic speed profile—flow diverted downwards over the sphere may create lift as well as drag. At right, the aerodynamics of the real tumbleweed involve flow through as well as around the structure.

It should also be noted that the inertia of a Tumbleweed may be significant, especially in the thin Martian atmosphere. Thus the rover does not follow the instantaneous wind. Inertia, together with ground friction, make the motion performance of the rover a nonlinear function of the windspeed history—it may be that a short gust gets the rover rolling, and in so doing contributes much more to the rover's distance on a given day than does all the rest of the time spent with weaker winds. Thus the windspeed distribution needs to be considered, and the high-end tail of the distribution in particular. It has been shown that Martian windspeeds, like those on Earth, can be fit with a Weibull distribution.

The question might arise of how the motion of a Tumbleweed might be directed. If the shape were altered, for example by deflating part of the vehicle, it is unlikely to have a long-term effect, causing it to move in an arc.

Significant deflation may allow a Tumbleweed to "rest" in one place, and this behavior could perhaps be programmed to occur at a certain time of day. Martian winds are in fact rather predictable; the thin atmosphere has little dynamic inertia or heat capacity, and thus little "memory." Winds near the surface tend to be dominated by slope winds—upslope during the day, and downslope at night. The latter are familiar on Earth as katabatic flows—these can be very strong on the margins of the Antarctic ice sheet, for example. They may be familiar to anyone who has camped in what looks like a pleasantly sheltered gully in mountainous terrain. Instead of a cosy shelter, the gully can be a conduit for a torrent of chilly air.

Thus, in an area with a known regional slope, it might be possible to "ratchet" the motion of the vehicle in at least two possible directions—either upslope or downslope—although with adequate weather measurements or models it may be possible to guide it with even more fidelity.

One test was performed from the National Science Foundation's Summit Camp site in Greenland in August 2003. This rover reached speeds (determined from an on-board Global Positioning System (GPS) receiver, relayed by an Iridium satellite phone and modem) of up to 16 km per hour. In its traverse lasteding 19 hours and 45 minutes, the rover dropped in altitude by 80 m but moved some 131 km from its release position. Its batteries permitted further transmissions for another 8 days, but evidently the rover had deflated or was lodged somewhere.

Figure 3.15. Prototype Tumbleweed in the lab just prior to tests in Barrow, Alaska. Notice the inflation port at the bottom, and the array of studs for wear resistance and perhaps boundary layer control. Photo NASA/JPL/Behar.

In the case of Mars, a 6 m diameter ball is easily capable of climbing over one meter rocks and up 25° hills (well over 99.9% of the Martian surface) with typical global winds that occur during the southern summer. The ball could also potentially be used as a parachute on Mars (30 m/s descent rate) and as an airbag, thus serving as its own landing system.

Various architectures have been explored, beyond the obvious inflatable sphere. In one desert test, an elongated tumbleweed was deployed with the expectation of a preferred rolling direction. Indeed, this sausage-shaped rover began to roll as expected, but as it acquired larger kinetic energy, bounces permitted it to make the energetically favorable transition to rotation about its maximum moment of inertia, and it began bouncing end over end instead of rolling. Although an interesting example of a basic mechanical principle discussed in many instances in this book, the behavior does not recommend this as an attractive vehicle design.

Other design approaches have considered noninflatable solutions—indeed in many respects a smooth ball is the least effective way of harnessing the wind. Some other approaches include using collapsible structures that spring out into a spherical shape. Also, rather than a smooth sphere, architectures like that of a dandelion flower (with stalks radiating from a central point tipped with flat ends that define a spherical surface), or a collection of conical cylinders (literally like polystyrene coffee cups) to make a very high drag object but with a small contact area. A study funded by the European Space Agency at the Helsinki University of Technology explored a 1.3 m diameter rover made from flexible 8 mm glass-fiber rods bowed around a central cylinder. The 18 rods each carried a "sail" made of nylon fabric.

So far the Tumbleweed concept, in any of its incarnations, has not reached sufficient technical maturity to fly. However, the progressive improvement and miniaturization of electronic systems, and the obvious logic of using Martian winds to traverse the surface, suggests that its chances can only get better.

References

Antol, J., P. Calhoun, J. Flick, G. A. Hajos, R. Kolacinski, D. Minton, R. Owens, and J. Parker, Low cost Mars surface exploration: The Mars Tumbleweed, *NASA/TM-2003* 212411, August 2003—see also http://centauri.larc.nasa.gov/tumbleweed/.

Behar, A., F. Carsey, J. Matthews, J. Jones, NASA/JPL Tumbleweed polar rover, *IEEE Aerospace Conference*, Big Sky Montana, March 2004—see also http://robotics.jpl.nasa.gov/~behar/JPLTumbleweed.html.

Bridge, N. J., The way balls bounce, *Physics Education* 33, 174–181, 1998.

Brody, H., Physics of the tennis racket, *American Journal of Physics*, 47, 482–487, 1979.

Brody, H., Physics of the tennis racket II: The "sweet spot," *American Journal of Physics*, 49, 816–819, 1981.

Carre, M. J., D. M. James, and S. J. Haake, Impact of a non-homogenous sphere on a rigid surface, *Proc. Inst. Mech. Engrs 218 Part C: Journal of Mechanical Engineering Science*, 273–281, 2004.

Chapman, A. E., and R. N. Zuyderhoff, Squash ball mechanics and implications for play, *Canadian Journal of Applied Sports Science* 11, 47–54, 1986.

Cross, R., Dynamic properties of tennis balls, *Sports Engineering* 2, 23–33, 1999.

Cross, R., Measurements of the horizontal coefficient of restitution for a superball and a tennis ball, *American Journal of Physics* 70, 482–489, 2002.

Cross, R., The bounce of a ball, *American Journal of Physics* 67, 222–227, 1999.

Garwin, R., Kinematics of an ultraelastic rough ball, *American Journal of Physics*, 37, 88–92, 1969.

Goodwill, S. R., and S. J. Haake, Ball spin generation for oblique impacts with a tennis racket, *Experimental Mechanics* 44, 195–206, 2004.

Gualtieri, M., T. Tokieda, L. Advis-Gaete, B. Carry, E. Reffet, and C. Guthmann, Golfer's dilemma, Subm to *American Journal of Physics*, in press.

Iwatsubo, T., S. Kawamura, K. Miyamoto, and T. Yamaguchi, Numerical analysis of golf club head and ball at various impact points, *Sports Engineering* 3, 195–204, 2000.

Janes, D. M., The Mars ball, A prototype Martian rover (AAS 87-272), *The Case for Mars III, Part II—AAS Science and Technology Series*, ed. Carol R. Stoker, Volume 75, pp. 569–574, 1989.

Jones, J. A., Inflatable robotics for planetary applications, *6th International Symposium on Artificial Intelligence, Robotics and Automation in Space I-SAIRAS*, Montreal, Canada, June 19–21, 2001.—see http://www2.jpl.nasa.gov/adv_tech/rovers/pub.htm.

Julien, P., *Erosion and Sedimentation*, Cambridge University Press, 1998.

Lorenz, R. D., J. Jones, and J. Wu, Mars magnetometry from a Tumbleweed rover, IEEEAC paper #1054, *IEEE Aerospace Conference*, Big Sky Montana, March 2003.

Matthews, J., Development of the Tumbleweed rover, JPL Report, May 2003.

Roberts, J. R., R. Jones, and S. J. Rothberg, Measurement of contact time in short duration sports ball impacts: An experimental method and correlation with the perceptions of elite golfers, *Sports Engineering* 4, 191–203, 2001.

Spencer, D. A., R. C. Blanchard, R. D. Braun, P. H. Kallemeyn, and S. W. Thurman, *Mars Pathfinder* entry, descent, and landing reconstruction, *Journal of Spacecraft and Rockets* 36, 357–366, 1998.

Sullivan, R. et al., Results of the imager for *Mars Pathfinder* windsock experiment, *Journal of Geophysical Research* Vol. 105, (E10), pp. 24,547–24,562, 2000.

Tokieda, T., J. Spheres, contact, *Pour La Science*, Dossier No. 41, 20–23, Oct–Dec 2003.

Waye, D. E., J. K. Cole, and T. P. Rivellini, *Mars Pathfinder* airbag impact attenuation system, *AIAA-95-1552-CP*.

4
Spinning Bullets, Bombs, and Rockets

Rifling

After realizing the dispersion on spherical projectiles that spin could produce, Robins determined that spin can *improve* the accuracy with which a bullet, and in particular a streamlined bullet, can be shot. By putting spiral grooves in the barrel—rifling—the bullet can be made to spin in the desired direction. Gunmakers developed their own favored patterns of rifling grooves in gun barrels—the number of grooves, their shape (the lumps between the grooves are referred to as "lands"), and the rate of twist being a matter of experience and style. It took some time, however, for rifling to be quantitatively understood.

Figure 4.1. Rifling in a 155 mm artillery barrel.

While the shape and number of grooves still remains something of a matter of convention, tightly coupled to the slight deformation of the bullet in the barrel, the spin rate, which relates to the groove twist, has a quantitative and generally agreed-upon basis. A convenient formula, widely used (perhaps sometimes beyond its strict realm of applicability) is the Greenhill formula, which says that the twist length in the barrel, measured in bullet calibers, should equal 150 divided by the bullet length in calibers. In other words, a 7.6 mm diameter bullet that is 20 mm long should twist in fifty calibers or less—the spiral of grooves should have a pitch of about 35 cm or less. (Alfred Greenhill was Professor of Mathematics at the Woolwich arsenal in London. He devised the formula in 1879, based on a series of accuracy tests.)

In modern times, ballistics is a deeply studied field and the search for performance by the military and manufacturers of ammunition has given rifling a high degree of sophistication. Formally, the tumble stability criterion relates to $(I_{xx}\omega)^2/C_{M\alpha}\rho I_{zz}V^2$, a term which should be large (>1) for stability—in other words high spin is good, low pitch moment. ($C_{M\alpha}$ is the derivative of pitch moment coefficient with angle of attack and is characteristic primarily of the nose shape. Sharper noses, while giving lower drag, tend to be less stable.) A slender bullet (with small I_{xx}/I_{zz}) will be harder to stabilize.

There is as always a tradeoff; a tighter rifling twist will improve stability, but exerts more stress on the bullet as it is fired, and will wear out more quickly.

Spin stability is, for bullets at least, not always a good thing, in that tumble may be desired in the terminal part of the flight, that through the target. Like the tumbling rugby ball in chapter 2 having higher drag, a long bullet that tumbles end over end will cause much more tissue damage than will a stable one that lances through. (Some bullets are designed with soft or hollow points such that they flatten on impact, forming a mushroom shape, with the same effect—larger cross-section and thus more rapid energy deposition in the target.) As an example, the bullet of the AK-47 Kalashnikov rifle was intended to tumble twice on passage through a 40 cm thick target.

Spin and Ballistics

While in direct fire—where the trajectory is short and fast enough to be fairly flat—a spin-stabilized projectile is ideal, this is not the case for all projectiles. In direct fire, the projectile will stay oriented by virtue of its gyroscopic stiffness in its launch direction which remains the flight direction, and thus the angle of attack remains small. However, in indirect fire (i.e., long-range artillery, where a projectile must be fired at an appreciable angle upwards to attain maximum range) the flight path angle varies throughout the flight by as much as ninety degrees.

On a flat planet with no air drag, the maximum range for a given launch speed is attained with an angle of elevation of 45 degrees. If the shot is long enough—as in some of the monster artillery of World War

1 such as the German "Paris Gun," which fired a 100+ kg projectile over some 130 km, reaching 40 km altitude—then the curvature of the Earth must be taken into account and the optimum angle will be a little lower.

In the presence of an atmosphere, the optimum angle will depend on the shape of the projectile, and specifically on its lift : drag ratio. An axisymmetric projectile will develop no lift at zero angle of attack, the condition in which it is launched. However, as gravity pulls the projectile down, the flight path angle will decrease to horizontal and so the angle of attack increases since a spinning projectile will still be in its original orientation of some tens of degrees upwards. Thus the projectile will develop some lift.

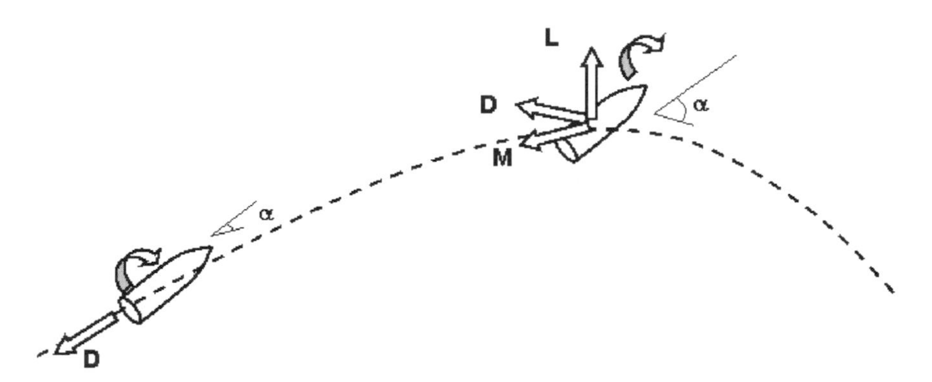

Figure 4.2. Schematic of shell in ballistic flight. At the beginning, the angle of attack α is small, and the only appreciable force is the drag *D*. Later in the flight, as gravity pulls the flight path angle down, α is higher and lift *L* and Robins–Magnus side-force *M* are developed.

Furthermore, because the projectile is spinning, it will develop a side-force from the Robins–Magnus effect (indeed, Magnus may have been motivated by a prize offered by the Berlin Academy of Sciences in 1794 to understand the deviation of artillery trajectories from predictions). This too needs to be taken into account in accurate long-range artillery fire. However, the side-forces experienced are not as straightforward as those discussed in the chapter on sports; slender artillery shells flying at supersonic and transonic speed can have complex shock wave and boundary layer separation effects. Wind tunnel tests show that the Robins–Magnus force for a given shell can change in sign depending on the flight speed, spin rate and angle of attack; things get even more complicated for finned projectiles (e.g., Regan, 1966).

Another spin that needs to be taken into account for very long-range artillery is that of the Earth—during the long flight-time the Earth's rotation can displace the target point by several kilometers relative to the free-flying shell. As a final complication, while conventional artillery shells are nice and solid, making their dynamics at least somewhat approximate to a rigid body, liquid payloads such as chemical weapons are sometimes considered. The sloshing and swirling motions of the liquid in a spinning shell pose new gyrodynamic stability challenges (see, e.g., Murphy, 1983).

Forensics and Rifling

Now popularized by TV series such as *CSI*, the imprint made on a bullet by the rifling grooves in the barrel during firing acts as a characteristic "fingerprint" of a given gun. Although in principle every example of a given mass-produced gun should have the same rifling, manufacturing tolerances and changes due to wear in use cause variations in the marks that get cut into the bullet. The marks allow bullets to be identified as being fired from the same gun.

Sometimes bullets are too deformed by impact to perform rifling analysis—one example here is a study to verify that a museum gun was indeed the one used by Booth to kill President Lincoln (Schell and Rosati, 2001). It had been alleged that the Deringer pistol in the Ford's Theater Museum which had thought to be the genuine item might instead have been replaced with a replica in the 1960s when museum security was less tight.

The Deringer pistol was rifled with seven grooves in a right twist. However, the bullet fragments removed from Lincoln's skull during the autopsy in 1865 had become too corroded for comparison, and so the study had to rely on defects in the barrel and comparison with photographs of the original weapon.

Figure 4.3. The Deringer pistol used to shoot Abraham Lincoln. Note the seven-grooved rifling pattern. Photo: Federal Bureau of Investigation.

Spinning Bombs

Some of the earliest air-dropped weapons were spin-stabilized. While earlier conflicts such as the American Civil War saw balloons used in reconnaissance, the First World War saw the use of aircraft as platforms for attack. The first bombs dropped from airplanes were improvised affairs, little more than encased explosive charges with fuses. It soon was realized that streamlining would permit bombs to fall more quickly (and thereby be less affected by wind). A series of bombs between 12.5 kg and 1000 kg developed by the German PuW (Prüfanstalt und Werft der Fliegertruppe = Test Establishment and Workshop of the Flying Corps) were sleek steel-cased weapons that were accurate in use from 1916.

The tail-fins of the PuW bomb were slightly canted, which caused the bomb to spin once released. This reduced the "wobble" amplitude, leading to a more consistent trajectory. A rather clever feature was that the fuze for the bomb was armed by centrifugal force—the bomb would only become "live" after reaching a certain spin rate, which would only occur after the bomb was released in flight.

In addition to improving accuracy, spin can improve bomb performance in other ways. Deep earth-penetrating weapons for attacking hard targets like bunkers or submarine pens rely on high impact velocity. While stabilizing fins at the back of a bomb can make the bomb statically stable (if displaced from zero angle of attack, it will tend to swing back), such oscillations may not be effectively damped, and thus the average angle of attack may be quite high. Spin can cause the average angle of attack to be much lower, and thus reduce the drag. One of the early applications of this approach was the "Grand Slam," a ten-ton "earthquake" bomb devised by British engineer Barnes Wallis in World War II. The improved stability allowed this 7 m long weapon to exceed the speed of sound in its freefall descent.

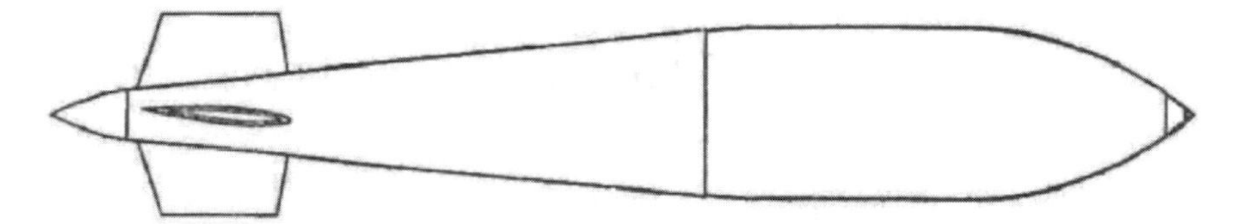

Figure 4.4. World War II earth-penetrating bomb (over 6 m long) with canted vanes to induce spin.

Another application of spin in bombs is in cluster bombs. These open up in flight to release many small bomblets or mines, for attacking convoys of vehicles, airfields and other area targets. Some simply rely on explosive charges to disperse the bomblets over a wide area, but some use spin. An example is the CBU-87 combined effects munition. This canister has fins at the rear whose angle can be adjusted to vary its spin rate after release. Once the desired release parameters are attained, the canister opens up to disperse some 202 small bomblets: the area over which these are dispersed depends on the spin rate, which can be set from 0 to 2500 rpm. At 1000 rpm the dispersal pattern is around 50 m across, whereas it reaches 80 m for spins of 2000 rpm.

More Spin on Weapons

Some bombs, while not themselves spinning, use a small "propeller" nut which becomes free to turn after bomb release—once the propeller has wound itself along a threaded rod during flight, it comes off and thereby arms the bomb. During the Falklands conflict in the early 1980s, some of this kind of bomb failed to detonate because of this fuzing system. Argentinian jets had to fly low to avoid anti-aircraft fire while attacking British ships, and the bomb flight time was so short that the fuze did not have time to arm. A number of ships were hit with bombs that tore straight through the hull, but failed to explode. The World War II German V1 flying bomb (essentially a cruise missile) similarly used the revolutions of a small propeller on the nose to determine that it had flown the ~200 km to its target.

Some guided weapons—both bombs and missiles—have been spun to facilitate their homing guidance. In early days, before sophisticated image processing with arrays of detectors became technically possible and affordable, there might only be a few sensor elements, just enough to see whether the target source (a laser beam bouncing from the target, or heat emitted from it) was one way or the other. To guide the weapon in both axes would require double the number of expensive sensors and actuators used to tilt the fins in response to the sensor signals. However, if the weapon was made to slowly spin, then the control authority in one body-fixed axis would be able to operate alternately in both directions—this approach is used by the Rolling Airframe Missile—a Sidewinder-derived shipborne anti-aircraft missile developed by the US Navy (e.g., Elko et al. 2001). Another approach is to spin a reticle (i.e., a coded mask) in front of the detector, so that the field of view of the detector is scanned across the field of view—the pattern of pulses from the detector allows, with knowledge of the reticle pattern, the position of the target to be identified and appropriate control signals generated.

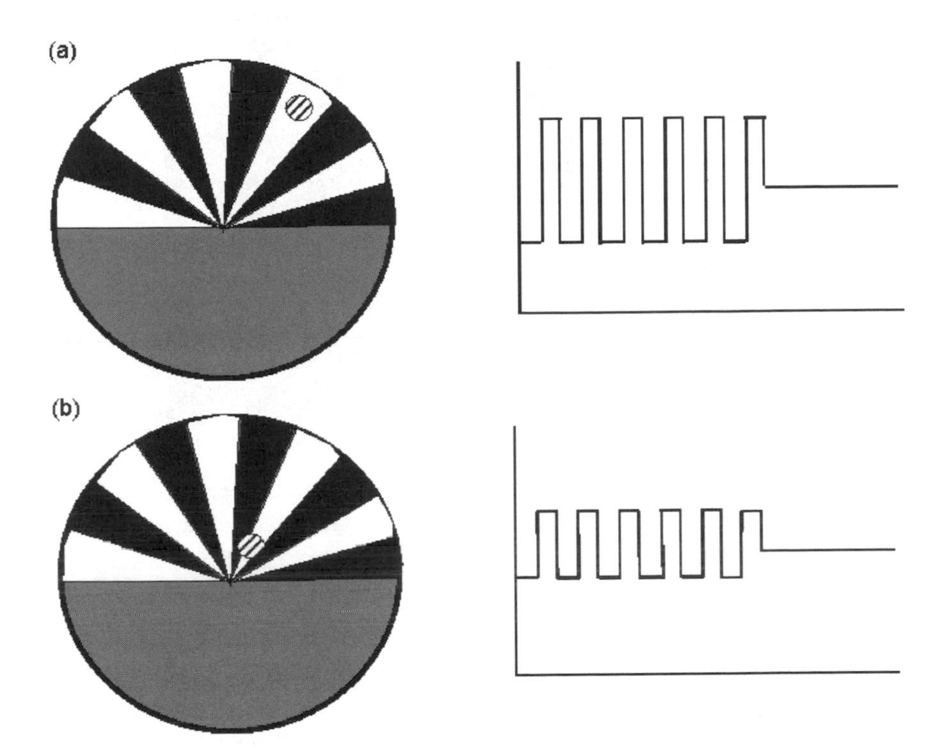

Figure 4.5. The signal from a target whose image (cross-hatched circle) is projected onto a spinning reticle will be modulated (right) in a manner that depends on the angle from the boresight—if far from boresight (a) the modulation by the radial pattern is stronger than at (b) where it is close to the boresight, allowing guidance signals to be generated.

A rather unique feature is found on the popular AIM-9 Sidewinder missile. This is an air-to-air heat-seeking missile used since the 1950s. The missile uses a spinning reticle infrared seeker on its nose to guide itself using steering fins (canards) at the front. The tail fins have small wheels at their tips which are spun up by the slipstream on launch. These wheels act to stabilize the missile in roll, and avoid the need for an active roll control system.

Figure 4.6. Tail fins of the Sidewinder air-to-air missile. Note the serrated wheel at the trailing edge tip—the serrations cause the wheel to be spun up by the airflow. Portion of USMC photo 200210135027.

Swords into Ploughshares

Not all guns are used in warfare. A series of experiments were conducted in the U.S. on high-velocity artillery to be used in a high-altitude research program (HARP) with shells making measurements in the upper atmosphere, much like sounding rockets. One of the main developers was Canadian ballistics expert Gerald Bull, who was some decades later involved in a plan to develop a long-range "Supergun" for Iraq. Instrumented shells were blasted to high altitudes (a 16-inch gun in Barbados fired a 185 lb projectile to some 140 km altitude (Murphy and Bull, 1966), and plans included rocket-assisted shells to attain even higher altitudes.

Interesting behavior can be seen in the attitude dynamics of projectiles. It was found that high altitudes could not be reached without canting the fins by a couple of degrees to induce spin (Mermagen, 1971). However, occasionally the spin period could resonate with yaw oscillations to cause discontinuous jumps in the spin rate (e.g., Figure 4.7).

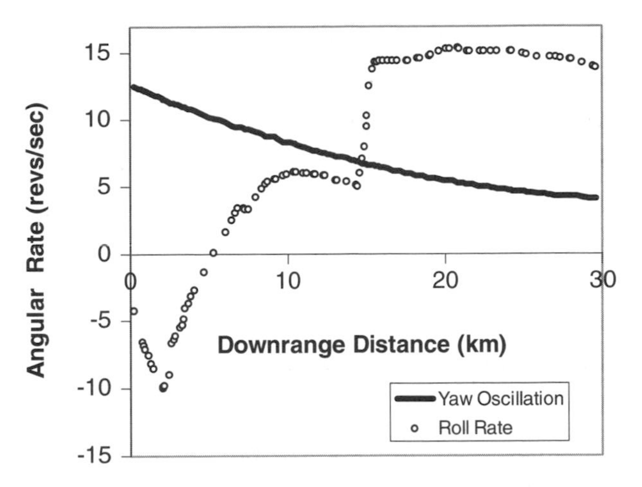

Figure 4.7. The roll rate in flight of a projectile from a HARP gun, measured with a sun sensor. The canted fins on the projectile seem to operate in reverse at first, then bring the projectile to around 5 revolutions per second. At that point, 15 km along the flight, the roll rate appears to couple into the yaw oscillation (the two motions have a similar period) and the roll rate is kicked up to 15 rps, from where it slowly declines for the rest of the flight. This roll–yaw coupling is a common concern for slender spin-stabilized vehicles.

Spinning Rockets

Spin is used to stabilize objects against small disturbing torques by momentum bias; having a large angular momentum vector in one direction will tend to keep the vector sum of that initial "bias" plus any additional angular momentum increments in the same direction.

There is another situation where spin is used for stability that has an entirely different principle. This is where spin is used to average out a misaligned thrust.

Rockets derive propulsive thrust by accelerating propellants in a nozzle. If, for example due to uneven combustion or a burnt-through nozzle, the thrust is misaligned, then two problems will occur. One, the vehicle will be accelerated in the wrong direction—for example, put into the wrong orbit. This problem may occur even if there is no change in attitude. The second, usually more common and more severe problem, can occur in the case above, or if the center of mass of the vehicle is displaced. If the thrust vector no longer passes through the center of mass, then it exerts a torque which tends to rotate the vehicle. A vehicle can be rapidly tumbled by a misaligned rocket thrust.

Now, if the rocket is spun around the desired thrust direction, a misaligned thrust will have an undesired sideways component (the sine of the misalignment angle) in a given direction. But half a rotation later, this sideways component will be pointing in the opposite direction and thus will cancel out the first part.

The First Intentionally Spin-Stabilized Rockets

Rockets, used by the Chinese for centuries before, became popular in the British military as an easily transported form of artillery in the Napoleonic wars beginning in 1805. The initial type was invented by William Congreve, and used a stick for stability. By moving the center of mass far behind the rocket nozzle, the rocket was stable. It was this type of rocket that was used in the War of 1812 with the new United States, most notably during the bombardment of Baltimore by British ships, providing the "rocket's red glare" that was immortalized in the U.S. national anthem. Both explosive and incendiary warheads were used.

However, the long stick made these rockets cumbersome to use and transport. Englishman William Hale devised a different approach in 1844: vanes in the rocket nozzle would cause the rocket to spin. By introducing spin early in the flight (or more specifically the "burn," that part of the flight when thrust is being applied), the thrust would be applied in a uniform direction, leading to more accurate flight.

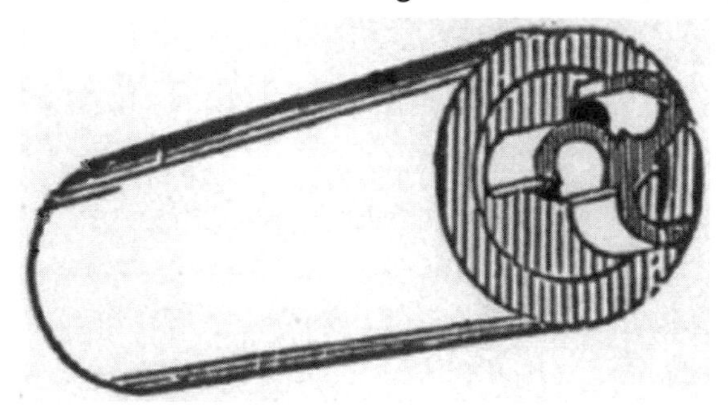

Figure 4.8. Curved vanes around the exhaust holes on the Hale rocket caused it to spin up—vanes as shown here cause the rocket to spin clockwise.

A popular size was the 24-pounder, light enough to carry in combat, with a range of up to 4000 m, although it was more typically used at 1200 m. The 23-inch long, 2.5-inch wide rocket was made of riveted iron sheet. The manufacturing approach was much more advanced than for the hand-made Congreve rockets—for example, a hydraulic press was used compress the gunpowder. Cylindrically curved vanes behind each of three exhaust holes at the base provided the spin. Grooves inside the casing ensured the powder propellant didn't shift during flight.

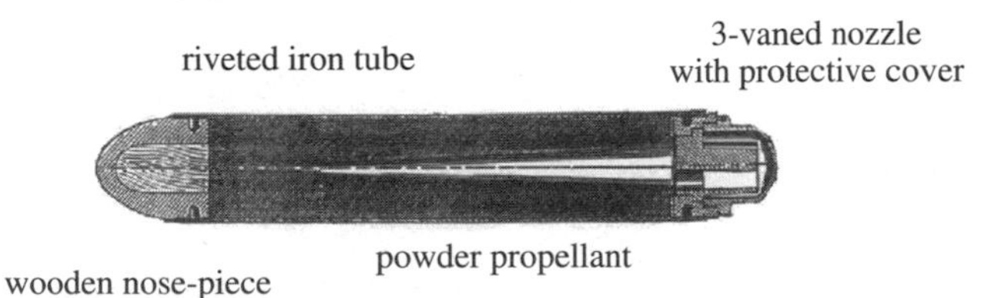

Figure 4.9. Cross-section of the compact Hale rocket—much easier to handle than a rocket with a stick. These projectiles were fired (after removing the end cap to protect against moisture) from tubes or open troughs.

Ironically, given the rocket's heritage in the Royal Army, Hale sold the manufacturing rights (he protected the design with several patents) to the U.S. government for some $20,000. The first use of Hale rockets was in the Mexican War of 1846. They were later used by Austria and from 1867 into the early twentieth century by the British in colonial conflicts in India and Africa.

20TH-CENTURY SPINNING ROCKET WEAPONS

Guided and unguided rockets—including sounding rockets for high-altitude research as well as missiles—continue to use various combinations of fin and spin stabilization. In some cases, the rockets have external canted fins to aerodynamically induce spin. In other cases, such as the 1950s-era artillery rocket Honest John (an 8 m long 2600 kg unguided rocket with a range of 25 km; the U.S.'s first nuclear-armed rocket), there are small rocket motors mounted circumferentially on the vehicle's body to rapidly cause spin as soon as it has left the launch rail. These spin motors give rise to dramatic exhaust plumes.

Figure 4.10. Honest John rocket leaving its launch rail. Bright plumes from just aft of the nose are two spin-up rockets. Photo: US Army Redstone Arsenal.

Spin-up rockets are used on more advanced missiles, too; the early 1970s saw the deployment of the MGM-52 Lance missile, which had a liquid-propellant motor and an inertial guidance system. This 6 m long, 1260 km weapon could throw a nuclear warhead some 120 km. It used 4 solid-fuel spin-up motors at launch, giving a characteristic black plume.

Figure 4.11. Lance battlefield missile leaving its transporter-launcher. Four black spiraling plumes from just aft of the nose are spin-up rockets. Photo: US Army Redstone Arsenal.

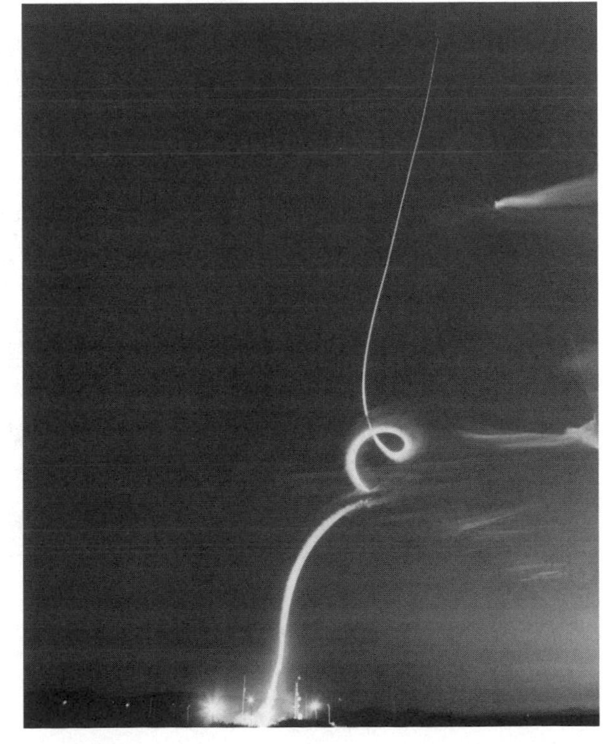

Figure 4.12. The spiral smoke trail of a Theater High Altitude Area Defense (THAAD) missile. The spiral trajectory is designed to bleed off speed in the confined test range and would probably not be used in anger. Lockheed Martin photo.

Even in these days of precision-guided munitions, unguided rockets are still big business—a General Dynamics contract for Hydra 70 2.75-inch rockets used by Cobra and Apache helicopters and the F-16 fighter in 1999 was over $1.2 billion.

Figure 4.13. Cobra helicopter fires a salvo of 2.75-inch rockets during an exercise. U.S. Marine Corps photo.

Figure 4.14. Ordnance crewman loading up a 2.75-inch rocket on an Apache helicopter during Operation Enduring Freedom in Afghanistan. U.S. Marine Corps photo 20046445947 by Cpl. Robert A. Sturkie.

These rockets have no guidance, and thus their accuracy is limited by their stability in flight and by the control exerted on their direction of flight before they are released. The rocket is fired from a tube, and while it is in the tube its pointing, and thus the direction of flight and thrust, can be controlled. Desirably, then, as much as possible of the impulse (the integral of thrust over time) will be exerted in the tube. In practice, there is a limit on how fast the propellant can burn, and so after initially accelerating out of the tube, thrust will continue for a short time. For this phase, a rapid spin is desirable to minimize effects of asymmetric thrust. The spin also helps to mitigate the effects of other perturbing forces, such as the rotor downwash when the rockets are used on helicopters.

On such a small and inexpensive vehicle as a 2.75-inch rocket, separate spin-up rockets are not practicable. Instead, a fluted exhaust nozzle—a little more sophisticated than Hale's, but the principle is essentially the same—is used to cause the rocket to spin up as soon as thrust begins to act. This provides the high spin-up torque needed to stabilize the rocket during its initial acceleration.

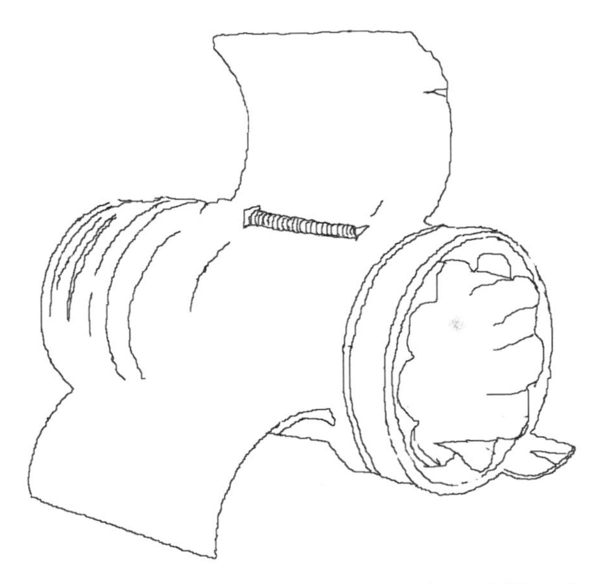

Figure 4.15. Sketch of the fluted exhaust nozzle for a 2.75-inch rocket, which causes the vehicle to spin up during launch. Curved fins which flip out after release cause the spin rate to decline again during flight.

However, the spin-up torque continues to act for as long as the motor is firing, and this leads to a fairly high spin rate during flight. The spin rate in flight should not be too high, since the bending mode of the long, slender projectile has a fairly low frequency. Just like a diver jumping up and down on a springboard, if the bending frequency of the rocket and the spin rate which will modulate any sideloads which might excite the bending mode are close in value, the rocket will resonate and become unstable in flight, and at the very least the accuracy of the rocket will be degraded.

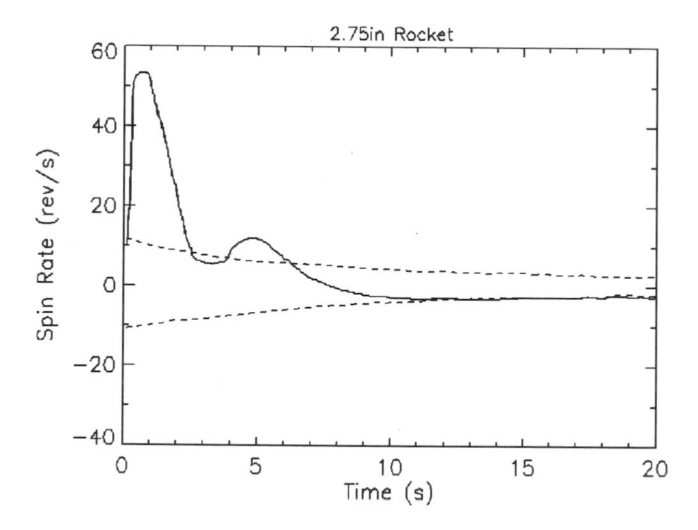

Figure 4.16. Spin rate of a 2.75-inch rocket (solid line). The initial spin-up is very fast, to stabilize the vehicle during the motor burn; the spin is rapidly reduced by the fins in order to get within the stable region bounded by the dashed lines indicating the bending frequency of the rocket structure. In this example, a small excursion is seen due to aerodynamic/bending interaction at around 4 seconds.

Thus the spin-up rate needs to be reduced as soon as possible after the motor has burned out, if not a little before. One approach used for many years is the deployment of small fins at the rear of the rocket (Bergbauer et al., 1980); in many instances these are curved fins which are mounted flush with the cylindrical wall of the rocket nozzle, and flip out when the rocket leaves its launch tube.

Another more recent approach is to construct the motor nozzle with internal vanes that cause spin-up as before, but to make the vanes from a material such as a plastic which erodes away in the hot gas efflux. By tuning the size, orientation, and composition of the nozzles, the amount of torque the vanes apply, and the length of time they apply it before they are dissolved away, can be manipulated. This allows the rocket motor to continue burning while no longer applying a spin-up torque.

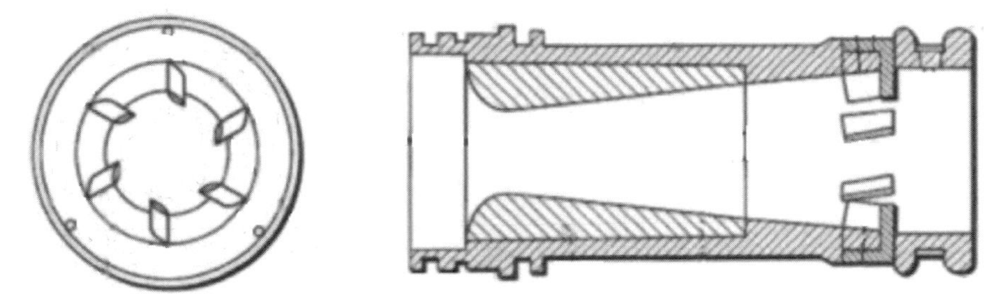

Figure 4.17. End-view and cross-section of the nozzle for a 2.75-inch rocket with erodable vanes, designed to operate only for the first part of the motor burn.

Sounding Rockets and Space Launchers

For the same reasons that military rockets are spun, sounding rockets for high-altitude research (often using the same hardware as missiles) are also spun at launch. The usual method used is to apply a slight cant to the tail fins of the rocket, such that an aerodynamic moment is applied on flight through the denser air at the beginning of the flight. (A related situation is discussed in chapter 7 on rotating parachutes in connection with the *Huygens* probe.)

The spin evolution may in this instance be somewhat complicated: the rocket accelerates faster and faster. Since the canted fins "demand" a spiral through the air with a certain number of revolutions per km, then the accelerating flight demands a faster spin rate. However, the air gets thinner and thinner as the rocket ascends, and so the torque falls — the demanded spin rate may not be attained in a time commensurate with the flight duration. On the other hand, since the rocket's mass and moment of inertia are decreasing rapidly as the propellant is consumed, it will accelerate more for a given torque. As with the 2.75-inch military rockets, care must be exercised to avoid spin–pitch coupling which can destabilize a rocket vehicle; underperformance of some rocket experiments has been attributed to high drag caused by the rocket pitching sideways due to spin coupling.

If the sounding rocket is a multistage vehicle, it may be that the spin-up is performed only while the first stage is attached. Often fins are only present—indeed, are only useful—on the first stage. Drag torques may slow the spin rate down at high altitude, especially if there are small uncanted fins on upper stages.

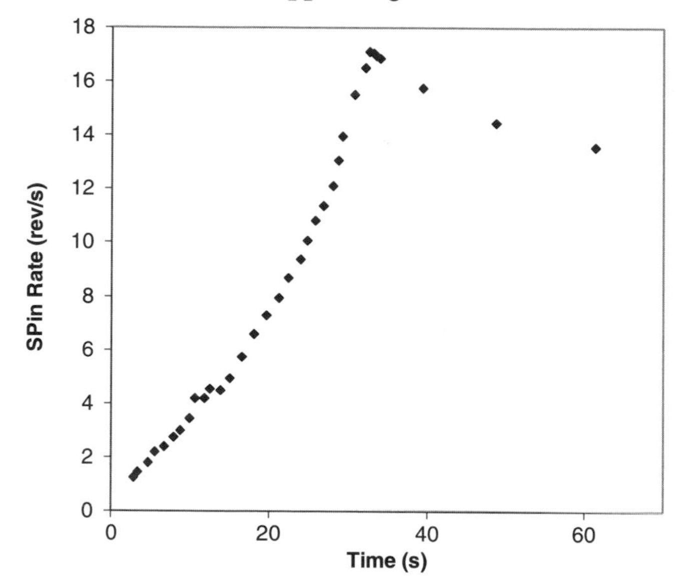

Figure 4.18. Spin rate profile of a Skua sounding rocket, as measured with an optical sun sensor (data from Williams, 1971). Note that the spin rate increases at an ever-faster rate as the rocket accelerates and burns up its propellant. After burnout the spin rate begins to decrease.

Sometimes gas jets or yo-yo despin devices are used to bring the spin rate to near-zero for the minutes or tens of minutes that sounding rockets spend at high altitude. Astronomical observation instruments such as UV or X-ray telescopes (which need to be above the atmosphere, which absorbs these wavelengths of radiation) usually prefer to stare at a target, which thus requires a low spin rate. Similarly, experiments in microgravity ("weightlessness") require low spin rates to avoid centripetal accelerations.

Occasionally, spin may be re-introduced on a sounding rocket to facilitate its intact recovery. The vehicle may experience aerodynamic heating on its re-entry into the Earth's atmosphere, although the heating at sub-orbital velocities is much less than from orbital or interplanetary speeds. Since the heating is modest, it may not be necessary to apply heat-shielding materials—the temperature rise on the rocket skin can be mitigated somewhat just by rotating the vehicle. Like a pig on a spit, rotation allows heating that is applied from a single direction to be applied uniformly over the whole circumference and thus prevent burning. Often payloads are recovered with parachutes; in cases where parachutes are not used the vehicle is often induced to tumble. This causes drag to be much higher (and thus the terminal velocity to be much lower and thus survivable) than if the rocket were to be allowed to fly nose-forward.

Large rocket launchers used to place satellites into orbit or on interplanetary trajectories generally do not spin, but rather use more sophisticated guidance systems and gimbaled rocket nozzles to correct for any thrust misalignments. However, upper stages used to reach a final orbit are often rather simple and use spin stabilization. This may be effected by a motorized spin table on the previous stage which spins the rocket up prior to separation. Sometimes separate spin jets are used.

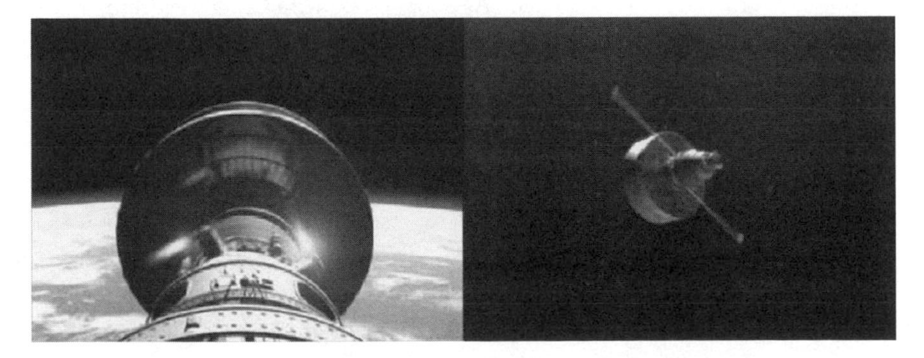

Figure 4.19. Excerpts from an animation of the launch of the Mars Exploration Rover (MER). At left the jets of two spin-up rockets can be seen, spinning up the MER vehicle (see chapter 7) and the third stage of the rocket booster. After the third stage has accelerated itself and MER onto a Mars-bound trajectory, yo-yo despin devices are unwound to reduce the spin rate.

References

Bergbauer, D. H., J. H. Ferguson, R. W. Bergman, and R. Bentley, Spin profile tailoring for the improved 2.75-inch rocket, *AIAA 80-1575*, Atmospheric Flight Mechanics Conference, 1980.

Elko, E., J. Howard, R. Kochansky, T. Nguyen, and W. Sanders, Rolling airframe missile: Development, test, evaluation and integration, *Johns Hopkins APL Technical Digest* 22, 573–581, 2001.

Mermagen, W., Measurements of the dynamical behavior of projectiles over long flight paths, *Journal of Spaceecraft and Rockets* 8, 380–385, 1971.

Murphy, C., Angular motion of a spinning projectile with a viscous liquid payload, *Journal of Guidance* 6, 280–286, 1983.

Murphy, C. M., and G. V. Bull, Review of the high altitude research program (HARP), *AGARD Conference Proceedings* No. 10, North Atlantic Treaty Organization, 1966.

Regan, F., Magnus effects, in the fluid dynamic aspects of ballistics, *AGARD Conference Proceedings* No. 10, North Atlantic Treaty Organization, 1966.

Schell, S. and C. Rosati, The Booth Deringer: Genuine artifact or replica? *Forensic Science Communications*, January 2001.

Stevens, F. L., T. J. On, and T. A. Clare, Wrap-around vs cruciform fins: Effects on rocket flight performance, *AIAA 74-777*, AIAA Mechanics and Control of Flight Conference, Anaheim, California, August 5–9, 1974.

Williams, E. R., A simple, reliable spin sensor for sounding rockets, *Journal of Physics E. Scientific Instruments* 4, 896–898, 1971.

Winter, F. H., The first Golden Age of rocketry—Congreve and Hale war rockets of the nineteenth century, *Smithsonian Institution Press*, Washington 1990.

http://www.redstone.army.mil/history/

5
Satellites and Spin

The dynamics of rotating bodies have been taken to extremes of precision and subtlety in the quiet of space. Entire books are devoted to spacecraft attitude dynamics and control; in this chapter we confine ourselves to some particularly illuminating or interesting examples.

Except for some applications (such as astronomy satellites) it is usually the case that a satellite maintains a constant orientation during its mission. This might be earth-pointing in the case of an observation satellite or a communications relay, or perhaps sun-pointing.

Spin also simplifies some elements of a satellite design. For example, if a box-shaped satellite which held a constant attitude in space were desired to maneuvre in a plane, it would need four thrusters in order to do so—forward and backward in each of two axes. However, if this maneuvre plane is the spin plane, a spinning satellite can maneuvre in all directions using only a single thruster by simply timing the thruster firing when the spin phase is correct.

Even for nominally nonspinning satellites, spin is also the usual condition in which the satellite begins its life, since its launcher or the upper stage used to deliver it to its final orbit is likely to be spin-stabilized.

Figure 5.1. A communications satellite (SBS-3) being deployed by spring from the cargo bay of the space shuttle *Columbia* in 1982. The satellite was spun on a turntable prior to release to keep it stable while the rocket motor at its base boosted it to its higher operational orbit. NASA Images S82-39791, S82-39793, S82-39794.

Studying Satellite Spin Dynamics

Spacecraft attitude dynamics are of course most accurately and conveniently studied simply by examining the attitude measurements made on-board by sun sensors, magnetometers and the like. However, this approach is of little use where no sensors are available or where the satellite is derelict or otherwise uncooperative. In this case, study of the electromagnetic radiation emitted or reflected by the satellite is the only avenue open for the investigator.

These remote studies were perfected to a high art in the early days of the space age, when attitude measurement and control was not yet developed. Special cameras for studying satellites and their reflected light were used to measure the orbits as well as the attitude variation, and radio telescopes and radars were frequently trained on satellites. It is easy enough, if you know where to look, to observe the flashes from spinning satellites and tumbling rocket stages with binoculars. Or, with a modern digital camera, you can record the streak of the satellite in the sky with a 15-second exposure and it may be possible to see the spin modulation of the light intensity. There is even an amateur group (PPAS, or Photometric Periods of Artificial Satellites, part of the Belgian Working Group Satellites) which maintains a database of flash periods; in addition to the steady despin by eddy currents (see below), occasionally rapid jumps in period occur due to venting or collisions.

The trick in observing satellites is knowing where and when to look. Prediction of passage of satellites or rockets overhead used to be a rather arcane business; the orbit could be calculated on a home computer easily enough, but because the orbits evolve due to thruster firings, the gravitational pull of the moon and the sun, the nonspherical gravitational field of the Earth, and (for low-orbiting satellites) atmospheric drag, these prediction programs needed to be regularly updated with revised orbital parameters. An orbit is defined by seven numbers—equivalent to a position and velocity in three dimensional space, plus the time ("epoch") to which they refer; for various historical reasons, the orbit is most commonly expressed in a different form, a set called "Keplerian elements" or "Keps" for short—the orbit radius, inclination, eccentricity, and so on. These in turn were often disseminated in an abbreviated form in the pre-Internet days as a three-line set of formatted numbers, called Three Line Elements or TLEs. Nowadays, it is easy to bypass that process and use a prediction tool on the Web. The most generally useful site is www.heavens-above.com.

Most satellites begin their operational lives in a state of high spin, since the rocket that deploys them into their final orbit needs to be spun to even out any thrust asymmetries. One of the earliest observations reported was the spin of the first satellite, *Sputnik 1*. The radio signal from this first satellite was heard by many all over the world. Dynamicist Ron Bracewell at Stanford measured the radio signal strength at a couple of wavelengths—when the variation was correlated between the two, it seemed reasonable to interpret this as a spin modulation. There were other uncorrelated variations that they attributed to ionospheric effects. Bracewell's collaborator in this work was Owen Garriott, who later was himself to become a satellite of the earth, floating inside the *Skylab* space station. He performed the graphic illustration of nutation damping discussed in chapter 1.

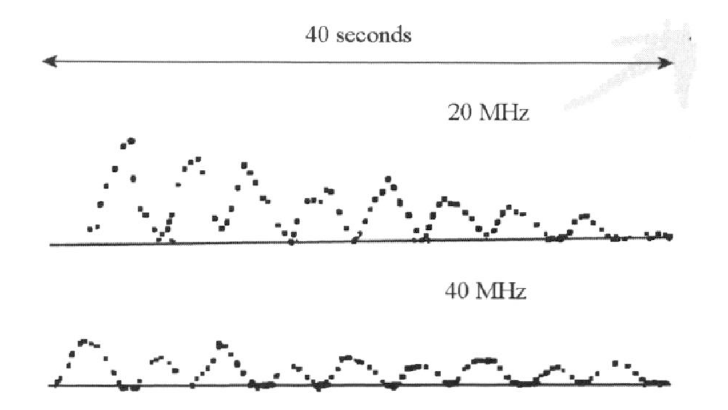

Figure 5.2. Bracewell and Garriott's record of the correlated fluctuations in the radio signal strength from *Sputnik 1*, recorded at Stanford on October 7, 1957. These indicate the spin of the satellite.

Before the days of home computers and orbit simulations for everyone, various analytic and graphical methods were used by those enthusiasts and professionals trying to track satellites. One of the most famous was an English high school physics teacher, Geoff Perry. Using some graph paper and a war-surplus radio receiver, Perry and his collaborators (known as "The Kettering Group") were able to deduce the existence (a state secret at the time) of a new high-latitude launch site in the Soviet Union, just by the timing of satellite passes overhead. In a patriotically inspired investigation, Perry, another teacher colleague, and Perry's daughter (a high school student at the time), measured the spin rate of the first and only British-launched satellite, *Prospero*, launched by the Black Arrow rocket from Australia in 1971 (Perry et al. 1973).

Radar is occasionally used to study satellites. When the European Space Agency's *SOHO* satellite went spinning out of control (Harland and Lorenz, 2005), the large radio telescope at Arecibo in Puerto Rico was used as a radar to diagnose *SOHO*'s spin state. From the strength of the reflection (a million and a half km away) astronomers could tell that the spacecraft was at an angle, and from the Doppler broadening of the echo they could tell it was spinning at one revolution per minute. (A narrowband radio signal will be broadened by reflection from an object rotating with its spin axis inclined to the view direction—some surfaces will be coming towards the radar and so blue-shifted or increased in frequency, whereas surfaces on the receding side will be red-shifted. The same technique was used to measure the rotation rate of Venus and Mercury.)

The light curve from satellites is dominated by specular (mirror-like) reflections from their many flat and shiny surfaces—flat antenna arrays, solar panels, mirrors for heat rejection, etc. Even relatively small surfaces can give strong reflections, but only in particular directions. At the other extreme, other objects such as asteroids (which we discuss in more detail in chapter 6) have rough surfaces which give a diffuse reflection. In this case, the brightness depends mostly on the projected area.

Another exotic case of a spinning space object is the inflatable geodetic satellite *PAGEOS* (Passive Geodetic Satellite). This was a 30 m diameter satellite, nominally a sphere, of aluminized mylar. If this satellite remained perfectly reflective and perfectly round, it would have no light curve at all. Its brightness as seen from the Earth would simply be as a result of reflecting an image of the Sun, as if the satellite were a slightly nonflat mirror. The (virtual) image of the Sun would be 7 cm in diameter. The image would appear at the specular point on the sphere, where the ray to the Sun, the ray to the observer, and the surface normal all lie in the same plane. Measurements of this brightness were made by ground-based telescopes equipped with photometers (e.g., Vandenburgh and Kissel, 1971).

Figure 5.3. The *PAGEOS* inflatable satellite during an inflation test inside a hangar on Earth. NSSDC Image.

During a typical pass over an observatory, the location of the specular point will move due to the changing geometry, describing an arc across the sphere's surface. The satellite had a rotation period of 190–280 s, and so during the visibility period of atypical pass of a few minutes, the satellite might rotate around six times. Although the location of the specular point drifted only slowly in this period, the satellite was rotating underneath it, and so the specular point (and thus the Sun's reflection) described a spiral pattern on the satellite's body.

The emergence, around 14 months after launch, of large brightness variations during these passes suggested that the satellite was no longer spherical. Specifically, the interpretation of the drops in brightness by about 3 stellar magnitudes (i.e., a factor of 16 or so in brightness) twice per rotation cycle was that the satellite had deformed into a prolate spheroid, like a rugby ball. Even though the surface reflection properties appeared from other measurements not to have changed substantially, such large drops in brightness could be generated by differences in the radius of curvature, which would be lower at the "pointy" ends of the satellite. In effect, these ends acted as mirrors with a shorter focal length, making the Sun's image smaller and so reducing the observed brightness of the reflection.

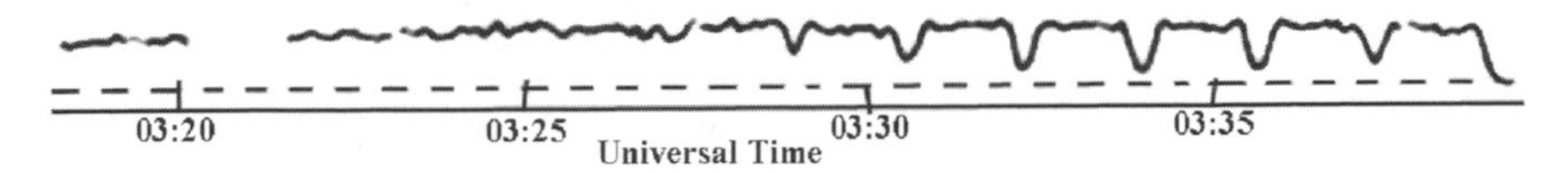

Figure 5.4. Curve showing the light intensity of the reflection of the Sun in the silvery balloon satellite *PAGEOS* during a pass overhead. The drops in intensity were interpreted as weaker reflections of the Sun due to the deformation of the satellite into a prolate shape. The dashed line shows the sky background level.

Spin Stabilization: Moments of Inertia

America's first satellite, *Explorer 1*, was essentially an instrumented upper stage of a rocket. It was therefore rocket- or pencil-shaped, and for stability (see sounding rockets) was spin-stabilized. It communicated with the ground by four whiplike wire antennae, forming a cross orthogonal to the spin axis. Although these wires could flex and therefore dissipate rotational kinetic energy, they were too short and low mass to significantly affect the moments of inertia. The spin axis was therefore the axis of minimum moment of inertia and so was not stable.

Not long after launch, the energy dissipation in the antennae had put the satellite from the intended 700 rpm spin into a 120 rpm flat spin—end over end. There is a tale (Likins, 1985) that Ron Bracewell had realized this would be a problem and tried to warn the engineers constructing the satellite that it would be unstable in its intended spin configuration, but Cold War secrecy prevented his message getting through.

Figure 5.5. A full-size model of *Explorer 1*. Note the stripes of paint to control its temperature, and the whip antennas. Left to right are JPL Director William Pickering, space scientist James Van Allen, and rocket pioneer Wernher von Braun. NASA Image.

Figure 5.6. The large drum-shaped Leasat military communications satellite was sized such that its diameter just fit in the space shuttle cargo bay. At left a Leasat is being deployed, "Frisbee-style"; at right astronaut Van Hoften is spinning a satellite up by hand after making repairs. NASA images.

Although conceptually a stable major-axis-rotating satellite can be a disk, in practice the shape (which generally can ignore aerodynamic considerations) can be much more exotic, and since even small masses can have important inertial effects if at large distances, slender booms are often encountered.

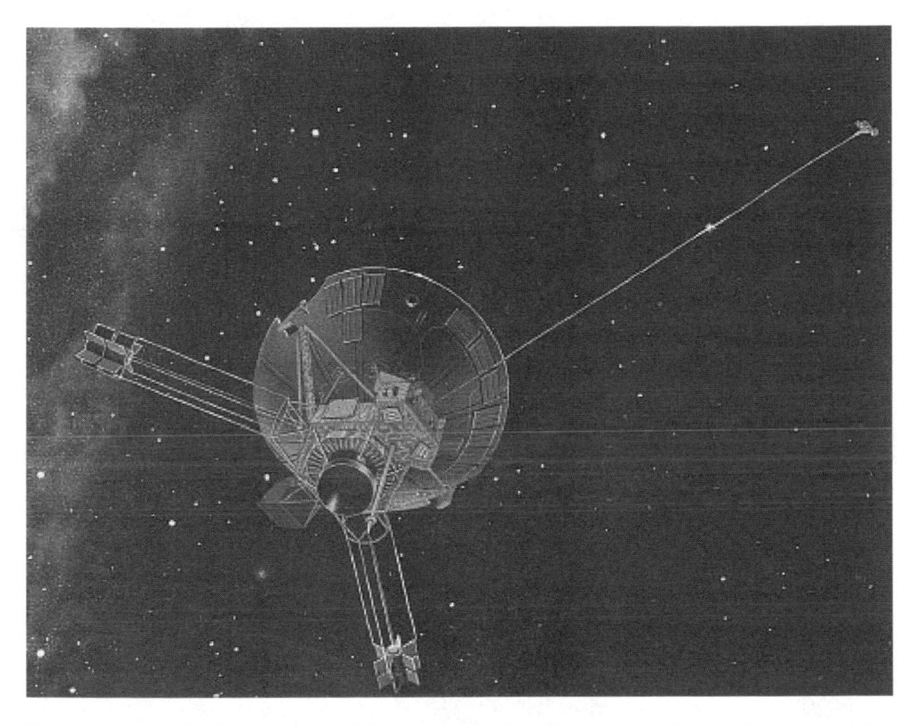

Figure 5.7. NASA's *Pioneer 10* spacecraft, the first to travel into the outer solar system. This satellite was spin-stabilized, keeping its high-gain dish antenna pointed towards Earth: the spin axis moment of inertia is kept high by two radioisotope generators held out on booms, balanced by a lighter but longer magnetometer boom.

Many satellites gain the benefit of gyroscopic stiffness without having any external part spinning at all. They carry a large flywheel or "momentum wheel," which adds momentum bias while allowing the rest of the satellite systems to be relatively fixed in space.

Another sophistication is the "dual spin" spacecraft, sometimes referred to as a "gyrostat" (e.g., Hughes, 2004). Here, part of the satellite is spinning for stability, usually a large cylindrical solar array, whereas part (for example the bit with the large Earth-pointing antenna on it) is despun. The stability considerations for these satellites are complex, and the spin bearing introduces a major complication, but this configuration is used in many geostationary telecommunication satellites.

Yo-Yo Weights

Usually large spin rates are only required early in a satellite's mission, for stability during a rocket motor firing to deliver it to its operational orbit. A satellite could then use thrusters or some other actuator, but this may be undesirably complex, slow, or expensive in terms of fuel.

One simple approach that is often employed is the use of "yo-yo" despin weights. These are masses (usually deployed in pairs) which are released from the body of the satellite on a tether that is wrapped around the satellite body. Centrifugal force holds the tether taut, and the moment of inertia of the satellite–weight system increases as if the weights were held out on rigid arms, causing the spin rate to drop, just as a skater holding her arms out will slow her spin. The angular momentum of the system remains the same as before.

However, at this point, the tether is cut and the weights fly off into space. (Another approach is that the tether is simply wound several times around the body of the satellite, if a suitable location for the tether is available—when the tether reaches its full extent, it just slips off.)

This leaves the satellite, minus weights, with a much lower spin rate than before. In essence, it has transferred some of its angular momentum to the weights. (This angular momentum transfer is effected by the tension on the tether acting at the rim of the satellite—hence the analogy to a yo-yo.)

The challenges of testing a yo-yo despin mechanism on Earth are formidable (e.g., Schiring et al. 1989). Although it is not too problematic to mount a satellite on a turntable, the extension of the yo-yo weights is limited by the free-fall height available to them, and in analyzing the results corrections have to be made for the air drag on both the weights themselves and the cable.

Gravity Gradient Stabilization

Beyond the intrinsic subtleties of spin dynamics, there are weak, but over long periods significant, torques exerted by the space environment. These must be taken into account, countered, or even exploited, for successful attitude control.

One rather deterministic (i.e., predictable) torque is that called the "gravity gradient." The gravitational acceleration caused by a spherical planet or a point source has a variation with distance. In other words, the center of gravity of an object (the "center of weight") is not in quite the same place as the center of mass. This means that the center of weight pulls down, exerting a torque until it, the center of mass, and the center of the planet are all in a line. Our own Moon is slightly bulged and this bulge points stably towards the Earth, meaning we see the same face of the Moon.

One stable situation is for the satellite to have its long axis always nadir-pointed (i.e., along the radius vector to the center of the Earth) and thus that long axis—the axis of minimum moment of inertia—rotates once per orbit in inertial space, about an axis parallel to the orbit normal. Thus the angular momentum vector, the axis of maximum moment of inertia, and the orbit normal are all parallel.

This approach, while mathematically sound and occasionally realizable in practice, is very sensitive to perturbations—a kick from a micrometeoroid, thermal flexing, or a misaligned thruster will set a satellite swinging. Indeed, a small amount of swinging (libration) occurs on the Moon. Gravity gradient stabilization is pendulum-like, with no intrinsic dissipation. A rigid satellite may therefore swing back and forth by large angles without settling down. There have even been cases where satellites have inverted themselves, becoming captured in a gravity gradient state but upside down.

Figure 5.8. The entirely passive *LDEF* (Long Duration Exposure Facility) being retrieved from orbit by the space shuttle's robotic arm. This NASA picture through the space shuttle's window shows *LDEF* in its stable Earth-pointed attitude. NASA Image EL-1994-00102.

One mitigating technique is to rotate around the long axis (i.e., to yaw), but to have this long axis rolled over slightly. The roll angle leads to a steady torque which causes the spin axis to precess. If the precession rate is fixed to the logical value of once per orbit, such that the long axis stays close to nadir-pointing throughout, then there is a one-to-one correspondence between the equilibrium roll angle and the yaw rate. For one satellite, *UoSAT-2*, that used this approach (Hodgart and Wright, 1987) the ratio of the pitch axis to the yaw (or spin) axis moments of inertia was made to be about 120 by extending a weight on a 7 m boom. With these moments of inertia, a roll bias of 5 degrees corresponds to a spin period of about 2 minutes.

The spin gives the attitude some stiffness, making the satellite less sensitive to perturbations. Another advantage is that the spin is favorable for evening out the amount of sunlight on different faces, evening out the temperatures—a "barbeque mode." However, occasional nudges are needed to this attitude and adjustments to the spin rate—these are accomplished by switching on electromagnets at calculated times (turning the satellite into a giant compass needle), commutating the current such that undesired motions are damped down. Achieving workable stability of 5 degrees or so with no thrusters or moving parts is quite a remarkable achievement.

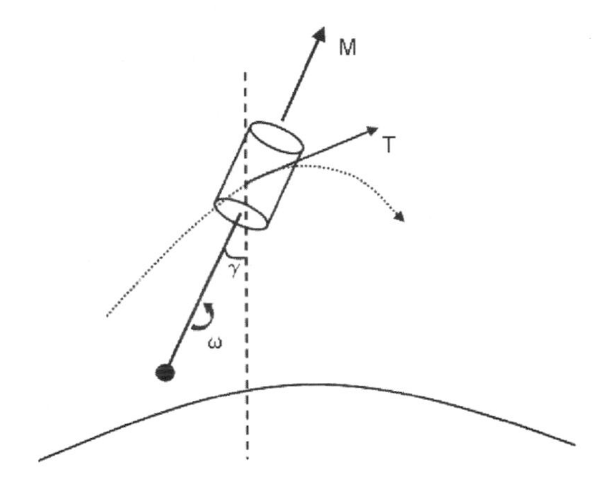

Figure 5.9. Schematic of a satellite in "barbeque mode" gravity gradient stabilization. Its long axis is rolled about the velocity direction (dotted line) at an angle γ from vertical. This causes a gravity gradient torque *T* which tries to swing the long axis back to vertical. However, the satellite is spinning about the long axis at a rate *w*, giving it an angular momentum *M*. The resultant effect of torque *T* is to precess *M* forwards, such that the spacecraft spin axis rotates forwards at the same rate as the orbital motion moves the "vertical" around in inertial space.

Magnetic Eddy Currents: LAGEOS Spindown

The magnetic field of the Earth affects satellite spin without even deliberate attempts to use a magnetic field on the satellite. Spindown of the satellites *Vanguard 1* and *2* were observed during their flights at the dawn of the space age, in 1958 and 1959. On *Vanguard 2,* the spin was observed over the satellite's three weeks of battery life to decay exponentially with a time constant of 72 days. *Vanguard 1* had a time constant three times larger—it was denser and thus had more moment of inertia and responded to torques more slowly.

The torque that slowed the satellite's rotation was eddy current drag: induction braking. A conductor moving in a magnetic field will develop a current which in turn generates a magnetic field which will act against the motion (this principle used to be used on speedometers for cars and bicycles—a magnet would be spun by a cable linked to one of the wheels, and would induce an eddy current torque on a metal disk. The disk would be held in place by a torsional spring, so that the angle of a needle attached to the disk would be proportional to the eddy torque, and thus to the magnet spin rate). Left to itself, a spinning conductive object in a field will therefore tend to spin down.

A rather nice illustration of this process is *LAGEOS* (Laser Geodynamics Satellite). This satellite is entirely passive, without power or transmission. It is in essence a mirror ball, designed so that laser beams can be efficiently reflected from it to precisely determine the distance between the ranging station and the satellite (similar reflectors were left by the Apollo astronauts on the moon). Made from the very dense material depleted uranium (to give the 411 kg satellite a very high ballistic coefficient so its orbit is not rapidly degraded by atmospheric drag or other perturbations), it carries 426 cube corner reflectors. *LAGEOS* observations show that the Pacific basin is moving to the northwest with respect to North America at the rate of about an inch and half per year—ranging measurements of the exquisite precision permitted by lasers permit direct measurement of continental drifts that otherwise had to be estimated less directly from the geological record.

Released into an orbit at around 5000 km altitude in 1976 with a spin period of about 1 s, the satellite has rather steadily spun down. Indeed, its spin period plots very nicely on a graph of logarithm of spin period vs. date (Figure 5.11), showing that the decay rate is proportional to the spin rate, and that the damping time for this satellite is of the order of 0.3 yrs. Another similar satellite, the Japanese Experimental Geodetic Payload (EGP, or "Ajisai") has many planar mirrors, rotates relatively rapidly and so sparkles when seen through binoculars.

Figure 5.10. The 0.6 m diameter *LAGEOS* satellite. Like a precision-engineered mirrorball it is covered with cube corner retroreflectors. NASA Image.

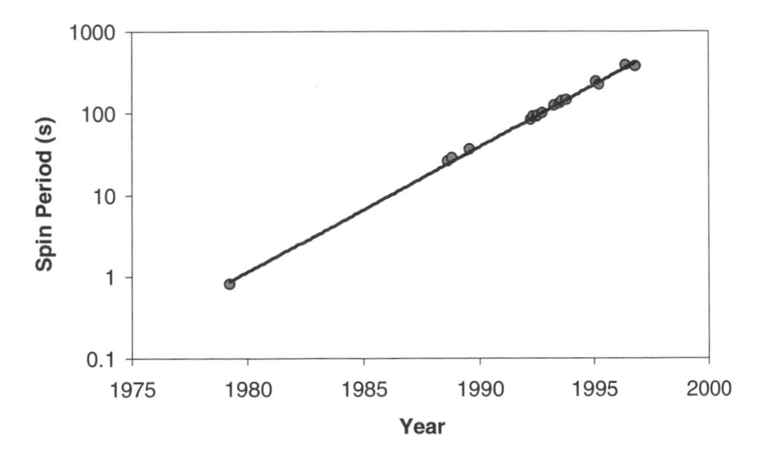

Figure 5.11. Spin period of the *LAGEOS I* satellite, showing the gradual increase of spin period as the satellite is slowed down. The exponential decay (indicated by a straight line of log spin period vs. time) is consistent with eddy current damping.

Solar Sailing

One torque that is sometimes seen, especially on satellites with large solar panels, is solar sailing. The light from the Sun exerts a tiny pressure on surfaces that absorb it at the Earth's distance from the Sun. The pressure is doubled if the light is reflected (since the photon's velocity is reversed, the change in the photon momentum and thus the momentum imparted to the mirror equals twice the original photon momentum). Large foil sheets have been proposed to propel spacecraft to other planets using this pressure—even though the accelerations that are attainable with thin aluminized plastic sheets are tiny, the pressure applies 24 hours a day, and can build up large speeds. A reflective solar sail spacecraft could be angled to "tack" at an angle to the sunlight pressure.

One design for a sail is the heligyro. This would use spin to unroll and hold rigid long thin blades as sails. Although the dynamics of such a vehicle would be complex, the spin obviates the need for extendable booms which would have to be prohibitively long. Less fancifully, the tiny pressure exerts a noticeable effect on spacecraft dynamics. If the spacecraft has asymmetric surfaces, the center of pressure will be offset from the center of mass, and thus the radiation pressure will exert a torque. The spacecraft *Mariner 4* used adjustable flaps at the end of its solar panels to control its attitude using this torque. In another example, geostationary communications satellites often have large flat solar arrays which can be tilted, and the small radiation pressure torque generated if these are tilted asymmetrically can be used to maintain attitude control (and thus permit the satellite to continue operating) after its fuel has been exhausted. One such satellite, *Olympus*, went into a spin after a failure in orbit (Harland and Lorenz, 2005), and could only be recovered after the Earth's orbit around the Sun moved sunlight relative to the spin plane and illuminated the solar panels adequately. The spin plane drifted slightly, changing the date at which this alignment occurred, due to solar radiation pressure on the canted solar panels.

A more subtle variation of solar radiation pressure torques was recognized in some early satellites with long boom antennae, specifically the Canadian *Alouette* (e.g., Hughes, 2004; Etkin and Hughes, 1967; Vigneron, 1973). Although the booms were symmetrical, and one would thus expect no net solar radiation pressure torque, a dynamical effect was found to occur, excited by the thermal distortion of the booms. In brief, on exposure to sunlight on the "dawn" side of the satellite, the sunny side of a boom warms and expands, causing the boom to bend slightly away from the Sun. This flexure in the boom exposes more projected area to the solar radiation pressure. Now, if the boom has a resonant period equal to some fraction of the satellite spin period, then the boom will be swinging through past its equilibrium (largest projected area) position as it rotates through the "dusk" side. Since the projected area is then lower on the dusk side than on the dawn side, the radiation pressure exerts a net torque to spin the satellite down. Similar effects were observed on other long-boomed satellites such as *ISIS* and *Explorer XX*.

As discussed in the next chapter, there can be radiation pressure torques due even to the heat radiated by objects which can, over millions of years, make major changes to the spin state of objects in space.

Docking: Matching Spin in Science Fiction and Science Fact

One of the classic moments in science fiction is in the film *2001*, when a shuttle from Earth arrives in orbit to dock with a space station. The space station, shaped like a giant wheel, is slowly spinning in order to produce artificial gravity. To dock smoothly, the shuttle positions itself along the spin axis of the station, and fires thrusters to make itself spin in inertial space at the same rate as the space station. Then, seen from the frame of reference of the shuttle, the space station is stationary, and the rest of the universe spins around them.

Figure 5.12. *Space Station 5* from Kubrick's classic movie *2001: A Space Odyssey*. The configuration has a stable set of moments of inertia—rotating like a wagonwheel. As the curved outer rim (the floor) rotates, it provides a centripetal acceleration to move the station's inhabitants in a circle, simulating gravity. The bright narrow rectangular aperture on the central axis is the hangar into which a shuttle from Earth would fly along the axis, after matching rotation with the station. NASA Image GPN-2003-00093.

In the same movie, the spaceship *Discovery*, which carries a crew to Jupiter, incorporates a spinning flywheel section, also for artificial gravity. In the sequel film, *2010*, this ship is now derelict. But the angular momentum of the flywheel has been transferred to the long ship, which is now in a flat spin. Space-walking astronauts must aim at the center of the ship, where the rotation causes least motion, in order to enter. These films are commendable in their accurate and silent treatment of spacecraft dynamics.

Although conceptually all that is needed to make artifical gravity is to spin such that the centripetal acceleration $\omega^2 R$ equals the desired "gravitational" acceleration, there is rather more to it. While $1g$ ($9.8\,\mathrm{ms}^{-2}$) of acceleration could be obtained with a 10 m radius space station spinning at 1 radian per second (10 rpm) or a 250 m station at 2 rpm. Obviously, the latter would be much harder to build. However, the first solution would be unworkable for humans—cross-coupling of head motions with the station rotation would cause motion sickness. Hall (1999) notes that while at 1 rpm no symptoms tend to be noticed, even pilots with robust tolerance to motion sickness could not adapt to a 10 rpm environment, even after 12 days. The size of the station can be made more manageable by allowing a lower gravity—perhaps $0.3g$ would be adequate to permit long-term health and permit normal walking and working.

Other fun effects would come into play—astronauts might be flung off a ladder by Coriolis forces as they ascend radially towards the rotation center. Perceived gravity would be more if walking along the curved floor of the station in the direction of rotation, and less while

Figure 5.13. The satellite *Westar 6* stranded in low orbit when its motor failed is recovered by shuttle astronaut Dale Gardner. Using a Manned Manoeuvring Unit (MMU) "jetpack," he inserted an apogee kick motor capture device (ACD) into the nozzle of the spent *Westar* engine to stabilize the satellite for capture by the shuttle arm and return to Earth. When the ACD locked Gardner to the satellite, its spin angular momentum was shared between them. NASA photo STS51A-46-057.

moving in the opposite sense. The artifical gravity has a gradient—things near the floor would fall faster than things at head height. Ball games would literally take an interesting turn.

Sadly the pace of development and use of large-scale space infrastructure is such that dealing with these challenges still seems some way off. Although the near-routine docking in space of capsules and space shuttles with the *International Space Station* (which does not spin) does require matching its dynamical state, the only events that really resemble the *2001* docking paradigm have been some in-orbit satellite repairs, where spacewalking astronauts have had to capture gently rotating satellites.

The Spin on *Gemini 8*

The Apollo strategy adopted by NASA to reach the moon required docking in space. The considerable complexity in such maneuvres was the price paid for using a much smaller rocket than would otherwise be needed. As a result, NASA undertook a precursor program, Gemini, to perfect docking techniques.

One of these missions, *Gemini 8*, flown by Neil Armstrong and Dave Scott in March 1966, nearly spun out of control. In addition to practicing docking maneuvres, Scott was to perform an extended spacewalk. After their launch, a day late, they rendezvoused successfully with an unmanned *Agena* target spacecraft. After inspecting it from a few meters away, they slowly closed with it and then attached—the first ever docking in space. However, soon they realized their spacecraft was rolling. Armstrong used the *Gemini*'s orbit and attitude thrusters to stop the roll, but it began again, spinning them faster and faster. Alert to the possibility the *Agena* might be misbehaving, they undocked, but the spin accelerated. An electrical short had caused one the orbital thrusters to stick. The crew switched off the orbital thruster system, and used a separate set of entry thrusters to bring the spin under control. They were safe, but their mission had to be aborted and they had to return early to Earth, to a secondary recovery area.

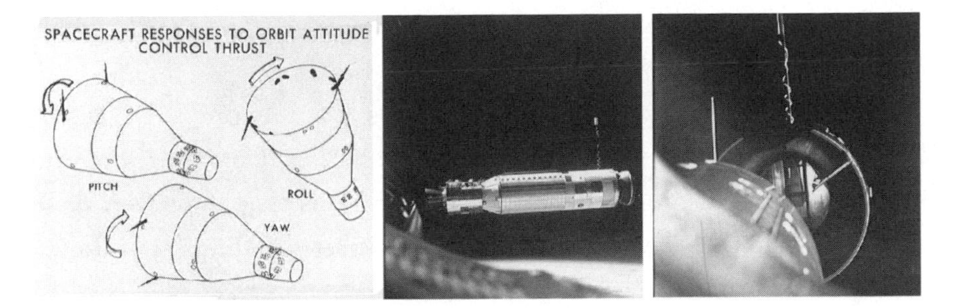

Figure 5.14. Pairs of thrusters of the *Gemini* capsule's orbit and attitude control thrusters were used to change its attitude. The capsule docked with the *Agena* target vehicle (center). Right panel is the view of the Agena from *Gemini*, just before its historic docking. NASA Images S64-03587, S66-25782, S66-25784.

Slosh

While we generally discuss spacecraft as rigid bodies, this isn't always the case. Occasionally spacecraft attitude dynamics appear externally rather erratic. This can occur if there are large bodies of liquid which are partly decoupled from the spacecraft body—the fuel and oxidizer for the rocket motor are often the culprits. An early example that was noted was the separation of the Apollo Service Module from the crew-carrying Command Module on return from the Moon. Although the separation was clean, the service module would often begin a puzzling motion afterwards. The design of propellant tanks must often take possible slosh into account and try to minimize it.

When space shuttle astronauts in 1992 recovered the largest commercial communications satellite (at the time) *Intelsat 603* for repair and relaunch (it was left stranded in a low orbit, and could be salvaged by attaching a new rocket motor), they were initially unsuccessful in getting hold of the satellite, which was spinning at a sedate half a revolution per minute. Indeed, after astronaut Thuot had attached a grapple fixture to the satellite, the satellite began spinning up again and the fixture came off. The propellant in *Intelsat*'s tanks had still been spinning around when the fixture was put on, and then transferred angular momentum back to the satellite body. This was an effect that had not been reproduced in underwater training on Earth.

A team of spacewalking astronauts was able to wrangle the satellite under control on the third attempt—it was still nutating slightly from the botched earlier attempts to grab it. This time, however, they held on to the satellite for several minutes to let the fluids spin down and settle before going on with the repair.

Figure 5.15. Astronauts on space shuttle *Endeavour* hold the giant *Intelsat 603* satellite while repairs begin. Plans for handling of the spin-stabilized satellite had to be hurriedly revised to take into account the motion of the fluid in its propellant tanks. NASA Image GPN-2000-10035.

References

Bracewell, R. N., and O. K. Garriott, Rotation of artificial earth satellites, *Nature* 182, 760–762, 1958.

Etkin, B., and P. C. Hughes, Explanation of the anomalous spin behavior of satellites with long, flexible antennae, *Journal of Spacecraft and Rockets* 4, 1139–1145, 1967.

Hall, T. W., Inhabiting artificial gravity, *AIAA 99-5424*, AIAA Space Technology Conference, Albuquerque, NM, September 1999.

Harland, D., and R. Lorenz, *Space Systems Failures*, Springer-Praxis, 2005.

Hodgart, M. S., and P. S. Wright, Attitude determination, control and stabilization of UoSAT-2, *Journal of the Institution of Electronic and Radio Engineers* 57 (supplement) S151–S162, September/October 1987.

Likins, P., Spacecraft attitude dynamics and control—A personal perspective on early developments, *Journal of Guidance* Vol. 9, No. 2, March–April 1986, pp. 129–134.

Perry, G. E., J. D. Slater, and I. J. Perry, Measurements of spin rate of the Prospero satellite, *Journal of the British Interplanetary Society* 26, 167–169, 1973.

Schiring, E. E., J. W. Heffel, and C. J. Litz, Simulation modeling and test of a satellite despin system, *AIAA-89-3267*, AIAA Flight Simulation Technologies Conference and Exhibit, Boston, MA, August 14–16, 1989.

Vanderburgh, R. C., and K. E. Kissell, Measurements of deformation and spin dynamics of the PAGEOS balloon-satellite by photoelectric photometry, *Planetary and Space Science* Vol. 19, p. 223, 1971.

Vigneron, F. R., Dynamics of Alouette *ISIS* satellites, *Astronautica Acta* Vol. 18, 201–213, 1973.

There are many excellent books on spacecraft attitude determination, dynamics and control, and many more general books on spacecraft systems engineering cover the subject in more or less depth. Wertz is justifiably considered "the bible" on the subject, but is expensive and becoming a little dated. I have found the book by Hughes to be particularly illustrative, but all are worth reading.

Hughes, P. C., *Spacecraft Attitude Dynamics*, Dover, 2004.

Thomson, T. W., *An Introduction to Space Dynamics*, Dover, 1986.

Wertz, J. R. (ed), *Spacecraft Attitude Determination and Control*. Reidel, Dordrecht, 1978.

6
Encountering Asteroids and Comets

Not only do artificial satellites spin in space, but so does everything else—stars, planets, comets, and asteroids. Since the latter are basically rigid bodies, their spin states are (usually) relatively straightforward. But the spin can have a complex history, reflecting how the asteroid was formed or whether it was hit by another asteroid or meteoroid, or the spin can be the result of more subtle effects. One interesting puzzle is that some families of asteroids (asteroids in similar orbits, suggesting they were once part of a larger whole) have spin properties that cluster together (Slivan, 2002).

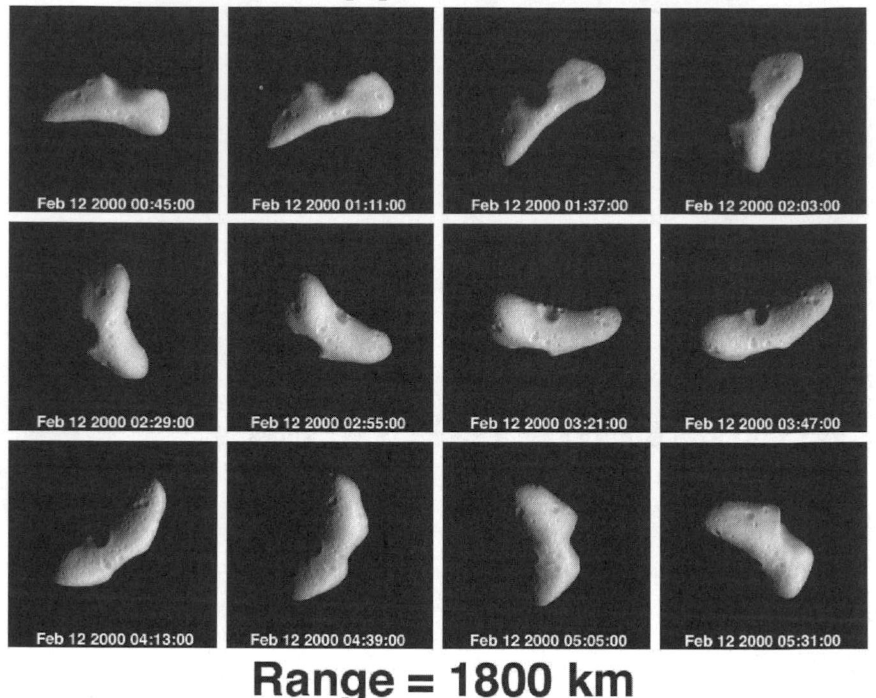

Figure 6.1. Rotation of the asteroid 433 Eros. The *Near Earth Asteroid Rendezvous* (*NEAR*) spacecraft, which later landed on Eros, took this 5-hour sequence of images looking down on Eros's north pole about 2 days before it entered orbit in February 2000. Image NASA-JHU/APL.

It is only in a few cases that the rotation of an asteroid is directly observed—where a spacecraft makes a close flyby of the asteroid, or when an asteroid makes a close flyby of Earth and it can be observed with radar.

More usually, the rotation period of an asteroid is determined by light curves. The asteroid appears only as an unresolved dot in the telescope. However, the brightness of the dot can be measured with precision (typically ~1%, sometimes much better). The brightness depends on the relative distances and angles between the Earth, asteroid, and Sun, factors which are usually known quite accurately, and which vary in a smooth, predictable fashion over days and weeks. However, the brightness is modulated, typically with periods of hours, by the body's rotation: if the asteroid is nonspherical in shape, and/or has areas of higher or lower reflectivity, then its effectiveness in reflecting sunlight into the telescope will vary as the asteroid rotates. (Asteroid light curves are usually smoother than those of artificial satellites, since asteroids generally reflect light diffusely, without strong glints.)

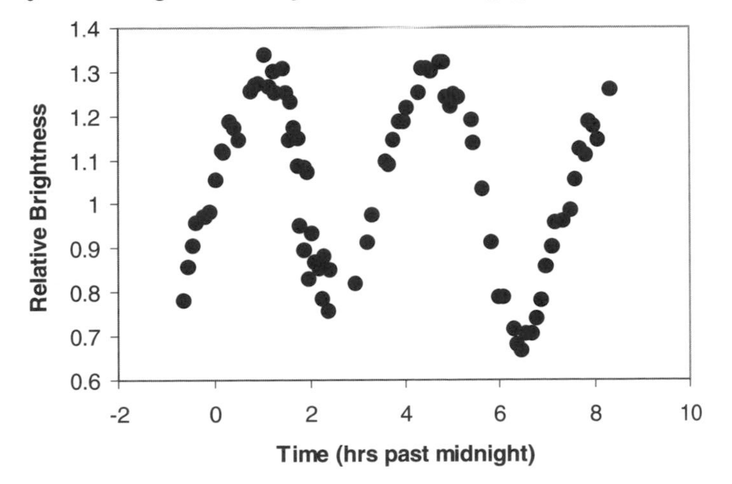

Figure 6.2. Light curve of the asteroid 311 Claudia, a member of the Koronis family (Slivan, 2002) showing a doubly periodic brightness during its 7.3-hour rotation period. In this view the amplitude is rather strong, about 30%—comparison with another lightcurve with a different geometry and amplitude allows the spin axis direction to be estimated.

Given a time-series of brightness measurements, the rotation period can be determined by a number of methods. One method, used especially by observers of fast rotators, is by simple "eyeballing"—plotting the brightness with time and seeing if an obvious periodicity emerges. This approach allows subsequent observing strategies to be fine-tuned, for example to make more or less frequent observations. Another obvious method is the Fourier Transform, useful when many different datasets are being compared together. Datasets over large periods can include spurious periodicities, notably the 24-hour diurnal cycle on Earth—astronomical observations are usually only acquired at night! Special algorithms are therefore invoked to filter out these spurious effects, CLEAN being one. When a correct rotation period is chosen, datapoints collapse onto a single curve of brightness versus phase (i.e., fraction of a rotation).

The asteroid 4179 Toutatis is a notable example where this collapse onto a simple curve does not happen. This asteroid is one of a population of asteroids whose orbits cross that of the Earth (the "Apollos"). It was discovered in 1989 by French astronomers, who named it after the deity in the *Asterix* comics. In December 1992 it passed within 0.0242 AU (4 million km), permitting close study. Observations from some 25 sites around the world were collated to estimate Toutatis's rotation period, but even when the rapidly changing viewing and illumination geometry were taken into account, it seemed that the light curve could not be fit with a single period (Spencer et al., 1986).

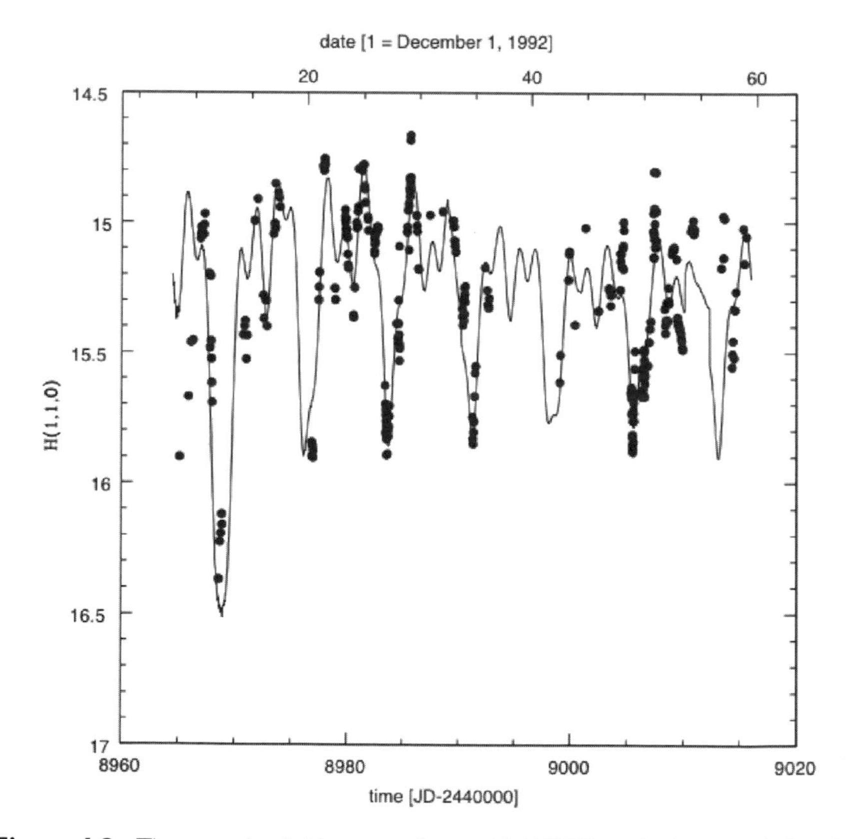

Figure 6.3. The complex light curve of asteroid 4179 Toutatis. Axes are Julian day (the graph spans about 2 months of observations) and magnitude normalized to 1 AU from Earth, 1 AU from the Sun at zero phase to compensate for viewing geometry. The curve is a complex rotator model by Mueller et al. (2002). Datapoints are from Spencer et al. (1986). Figure courtesy of Beatrice Mueller.

Images of Toutatis were synthesized using radar at the Deep Space Network site in Goldstone, California. A 400,000-watt coded radio transmission was beamed at Toutatis from the Goldstone main 70-meter dish antenna. Echoes were received, 24 or more seconds later, by a 34-meter dish and were decoded and processed into images. The coding allows different parts of the echo to be isolated, longer-delayed echoes coming from more distant parts of the asteroid, with the echo also spread in frequency by the Doppler shift; parts on the edge rotating towards the observer are shifted to a higher frequency. Extensive computer processing maps these echoes into a shape model.

Remarkably, its spin vector traces a curve around the asteroid's surface once every 5.41 days. During this time the object rotates once about its long axis, and every 7.35 days, on average, the long axis precesses about the angular momentum vector. The combination of these motions with different periods gives Toutatis its bizarre "tumbling" rotation.

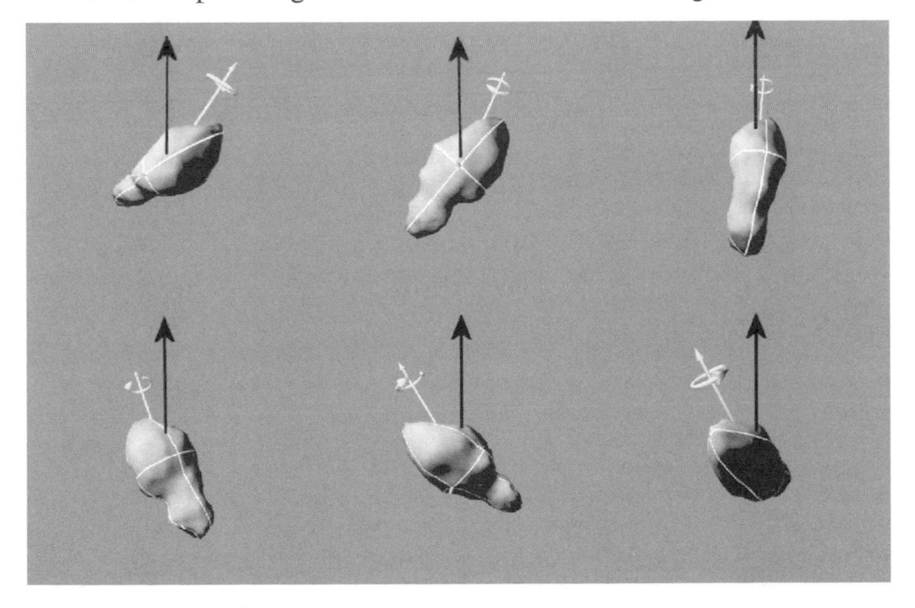

Figure 6.4. This image shows the non-principal-axis spin state of asteroid 4179 Toutatis at one-day intervals (read from left to right, top to bottom). The lines on the surface denote the principal axes of inertia; the constant vertical arrow is the angular momentum vector; the white arrow is the instantaneous spin vector with the sense of rotation shown. Unlike the vast majority of solar system objects that have been studied, Toutatis does not spin about a single axis; that is, it has no fixed north and south pole.

As mentioned in chapter 1, the "natural" state, in the sense of energetic stability, of a rotating object in free space is rotation about a principal axis. Usually this means the axis of maximum moment of inertia. If, however, the angular momentum vector is not aligned with a principal axis, then the object will exhibit an apparently complex motion.

Because such a motion is not simple to describe, it is sometimes termed "complex rotation." Another term is an "excited spin state," in the sense that energy dissipations will ultimately damp down the nutation, and thus the state must have been recently excited by some external stimulus. The same type of signal is evident in the motion of a Frisbee shortly after it is (badly) thrown. Because the state needs excitation on a timescale shorter than that needed to damp the nutation, it is relatively rarely observed on asteroids.

Asteroid Strength and Spin

Whirl a weight on a string around your head, and you will feel the string trying to slip through your fingers. Your hand pulls on the string to keep the weight moving in a circle. Similarly, the equatorial surface of spinning asteroid must experience a centripetal acceleration to keep moving in a circle. The source of the force causing this acceleration must either be the strength of the asteroidal material, if it is a solid lump of rock, or the gravity of the asteroid.

A plot of asteroid size vs. spin period is generally bounded by this limit (e.g., Pravec and Harris, 2000), suggesting that most large asteroids are simply self-gravitating aggregates of smaller elements—"rubble piles." While you can find small asteroids and big asteroids, and you can find fast rotators and slow rotators, you never find large, fast rotators.

Some smaller asteroids seem to beat the self-gravity limit, spinning so fast that they would tear themselves apart if they were rubble piles. This suggests these may be competent pieces of rock, perhaps collisional shards of once larger objects that were large enough to melt together. It is remarkable that such profound insights into the makeup of solar

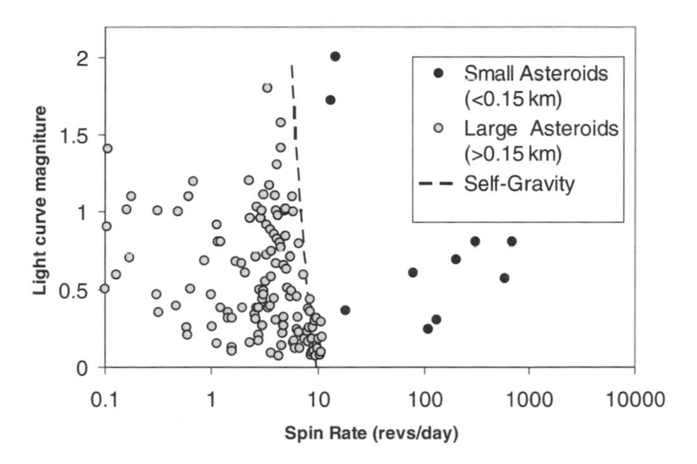

Figure 6.5. A plot of the spin periods of asteroids. Large asteroids—presumably rubble piles—seem to hit a "barrier" at about 12 revolutions per day, whereas small asteroids, presumably single shards of rock, can hold themselves together while spinning faster. The self-gravity barrier has a slight curve—more elongated asteroids with higher light curve magnitudes are more easily torn apart and so have a slower tolerable spin rate.

system bodies can be concluded from something as indirect as the measurement of spin periods.

Asteroid Spin and the End of the World

A perennial subject for science fiction movies is the impact of an asteroid or comet on the Earth, and the consequent mayhem. It has happened before—after all, the end of the hegemony of the dinosaurs over the planet was brought about (or at least hastened) by the impact of a 10 km asteroid into the Yucatan 65 million years ago. After long controversy, the impact and its effects are now generally accepted—and in fact the dinosaurs were particularly unlucky. While a 10 km asteroid, making a crater over 100 km across, is never trivial, the environmental effects of the Chicxulub impact were particularly severe because of the sulphate rocks that were vaporized by the impact, leading to a long "nuclear winter" and acid rain (which destroyed many sea species by dissolving their shells) to add to the global wildfires and local shock effects.

In any case, wherever it hit, such an impact today would be catastrophic for global civilization. Fortunately impacts of this size are only expected once in a hundred million years or so. But smaller impacts will occur much more often.

A small asteroid, perhaps 60 m across, exploded in the air above the remote Siberian forest near Tunguska in 1908. Trees were flattened out to about 15 km, and fatal injuries were received by reindeer herders 30 km away—the energy of the explosion, due simply to the kinetic energy of the incoming asteroid, was estimated at 2×10^{15} J, or about 0.5 megatons. We would expect an impact like this roughly once per hundred years—were this to happen over a city, the local death toll and the economic impact worldwide would be catastrophic.

Accordingly, governments are investing at least some effort in monitoring the asteroid threat. A number of telescopes around the world are dedicated to detecting and tracking near-Earth asteroids, and every so often an object is identified that has an orbit that might come close to the Earth. Especially early on, when only a few observations are available with which to estimate the orbit, the miss distance (from the center of the Earth—the miss distance may or may not be smaller than the radius of the Earth!) is poorly known. However, even with more observations to refine the orbit as it is now, there remain irreducible uncertainties in where the asteroid will be.

Perhaps surprisingly, the greatest source of uncertainty in assessing whether an observed asteroid will hit the Earth a decade or two from now relates to its spin. The spin can affect the orbit of an asteroid in a rather subtle way, known as the Yarkovsky effect after a Russian civil engineer who identified the effect at the end of the 1800s.

Every photon of light (or infrared radiation, etc.) has a tiny amount of momentum. If the photon is absorbed, or reflected, by a surface, then the surface must absorb, or reverse respectively, the momentum of the photon, and in so doing receives the momentum from it (or double the momentum, in the case of reflection). Thus a surface exposed to the Sun experiences a momentum flux, a pressure, from nothing more than the light. The force thus produced is tiny—an absorber at Earth's distance from the Sun experiences a pressure of only 5 micro-Newtons per square meter. This is all but insignificant for most objects, except those with exceptionally high area-to-mass ratios. Such objects include the so-called beta meteoroids (very tiny dust particles, observed to be streaming away from the Sun, pushed by radiation pressure) and "solar sails," an idea for interplanetary spacecraft that might harness the tiny pressure with enormous, lightweight reflective films. Some satellites have used their solar panels as improvised solar sails to perform small orbit adjustments.

But just as the short-wavelength, high-energy photons of sunlight exert a radiation pressure, so do the infrared photons associated with thermal emission. Launching the photons exerts a small pressure, just as absorbing them does. So a hot surface experiences a small pressure, but a slightly larger pressure than a merely warm one.

Hence, Yarkovsky realized, a body with an uneven temperature distribution would experience an uneven pressure, and hence a net force in space. A nonrotating body would of course be hottest on its sunlit side, and therefore would experience a net force away from the Sun due to the flux of thermal photons from that side (in addition to, but rather less than, the force due to the absorption of solar photons). However, a rotating body will smear the noontime peak in temperature around to the late afternoon, such that the dusk side of the object experiences stronger pressure than the dawn side, and there will be a net thrust towards the dawn side (see Figure 6.6). If the asteroid rotates very quickly (where "quickly" is defined by the thermal properties of the asteroid and how much sunlight it is exposed to), then the daytime heat bulge is smeared out completely, the temperature is uniform as if the asteroid were being roasted on a spit, and there is no net thrust. The process can even be considered as if the rotating asteroid were an engine, operating on the heat transported from the morning to the evening side—the engine has an optimum output power for intermediate rotation rates (Lorenz and Spitale, 2004).

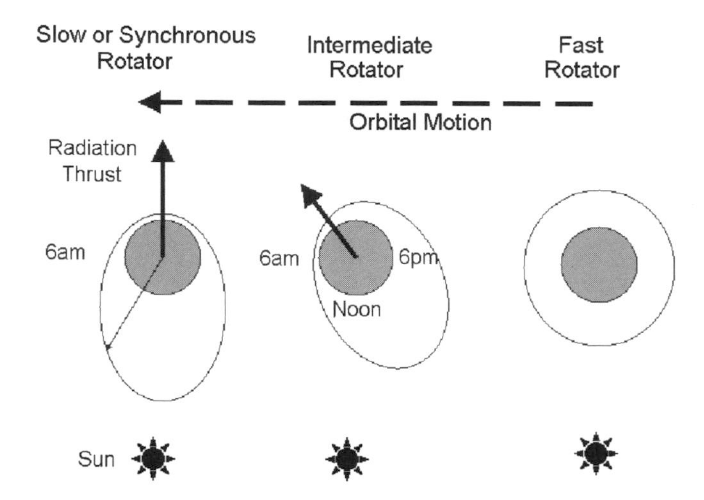

Figure 6.6. Schematic showing the radiation pressure distribution from a cylindrical asteroid (grey circle) viewed from its north pole. The ellipse is a polar plot of temperature, with a peak around noon for the slow rotator, giving a net radiation thrust away from the Sun. However, since this is orthogonal to the direction of orbital motion, no net work is done and the change to the orbit is tiny. A fast rotator has an even temperature distribution and so has no net thrust. An intermediate rotator has the temperature peak in late afternoon, with a modest net thrust that has a component along the direction of orbital motion which can significantly affect the asteroid's position in the future.

Operating over decades, this small force is enough to move asteroids measurably—precise radar measurements (Chesley et al., 2003) in 1991, 1995, and 1999 showe that asteroid 6489 Golevka had moved by 15 km from the position expected if the Yarkovsky effect were not acting.

There are variations on the Yarkovsky effect, depending on the eccentricity and inclination of the orbit, the angle between the rotation axis of the asteroids and its orbital axis, and the shape of the asteroid. These variations can modify the spin rate (e.g., Rubincam, 2000) as well as the orbit. The study of Yarkovsky effects and spin more generally is presently a major field of asteroid research.

Comet Spins

In a morphological sense, when you strip away the bright nebulous coma and the tail from a comet, what is left looks a lot like an asteroid—a giant potato in space, with perhaps a few craters on it. However, comets and asteroids have very different histories. While asteroids are rocks (or, as we just saw, piles of rocks), comets formed in much cooler conditions, and incorporate more volatile materials—organics, water ice, and volatile gasses in frozen form. When comets are pushed into orbits that plunge towards the inner solar system, the Sun's heat boils away these volatile materials, forming jets and a tail. The jets can significantly modify a comet's trajectory (astrodynamicists refer to "nongravitational forces"), making comet orbits difficult to predict more than an orbit in advance. The jets can also of course modify the spin of the comet.

The jets can also reveal the spin, in that the speed of the gas and dust emerging from the comet can be measured or modeled, and the shape described in space can be seen in telescopic images for comets that pass close to the Earth. If the comet is spinning with its spin axis close

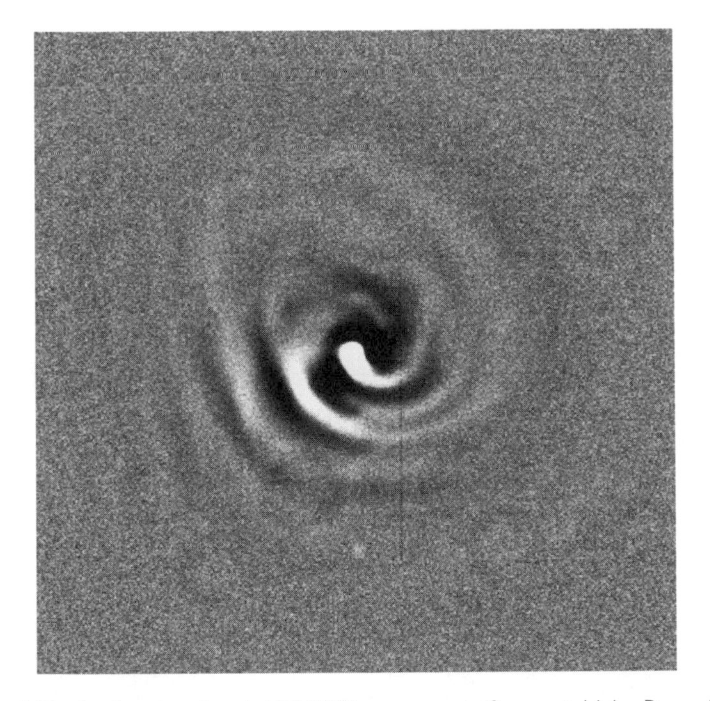

Figure 6.7. An image, about 100,000 km across, of comet Hale–Bopp in April 1996 showing the spiral pattern traced by the comet jets. WIYN telescope image provided courtesy of the WIYN Consortium, Inc. (Wisconsin, Indiana, and Yale Universities and the National Optical Astronomy Observatories) with support of the National Science Foundation.

to the line of sight, the jet will be sheared out into a spiral, like the exhaust from a pinwheel rocket. One good example was comet Hale–Bopp in 1996 (which was very visible to my naked eye, even from an illuminated sportsfield when playing Ultimate Frisbee—see chapter 8).

An analysis of the spirals and shells thrown out from comet Halley in 1986 (Belton et al., 1986) suggested it had a spin period of 3.69 days about its angular momentum vector, with the vector inclined to the long axis of the comet by 66 degrees (in other words, the comet is in an excited state, or non-principal axis rotation). The model yielded a ratio of moments of inertia of 2.28, and suggested that the comet density was more or less uniform.

Interpreting comet spin states from light curve data is particularly challenging, since light comes from the coma as well as the nucleus. For example, the light curve of comet Encke (Belton et al., 2005) shows some seven different periodicities at different times, suggesting an excited spin state coupled with variations in the active areas on the comet nucleus's surface. Although the spin dynamics of comets are interesting, if not bewilderingly complex, the rest of this chapter will concentrate on the somewhat more tractable problem of the attitude dynamics of spinning spacecraft when they encounter comets. These dynamics allow inferences about the dust spraying out from the comet.

Giotto at Halley

The first close encounter with comet nucleus marked the interplanetary debut of the European Space Agency, ESA. Early plans for a multinational mission to visit comet Halley in 1986 had initially faltered when the U.S. government failed to fund NASA's participation. Nonetheless, a small flotilla of spacecraft from around the world met Halley—a pair of Russian spacecraft, *VEGA 1* and *2*, led the way, after first visiting Venus, and two Japanese spacecraft made distant flybys of Halley. But ESA's *Giotto*, named after the Italian painter who included the comet in his painting *Adoration of the Magi*, was to brave the closest encounter, aimed a mere 596 km from Halley's nucleus.

The *Giotto* spacecraft was designed with the expectation that it might not survive the encounter with Halley's comet. The spacecraft would be flying around the Sun in an orbit not too different from the Earth's, while Halley was whipping around the Sun in the opposite direction on the closest, fastest part of its long, 76-year orbit. The combined collision speed was therefore some 68.373 km/s. At these speeds, even a speck of dust—and it is tons of such dust that makes up a comet's tail—has the kinetic energy of a rifle bullet. *Giotto* therefore transmitted all its data in real-time, without on-board storage, so that it would send as many of Halley's secrets as it could before it was felled. The datalink to ground was through an offset-fed dish antenna pointing back towards Earth.

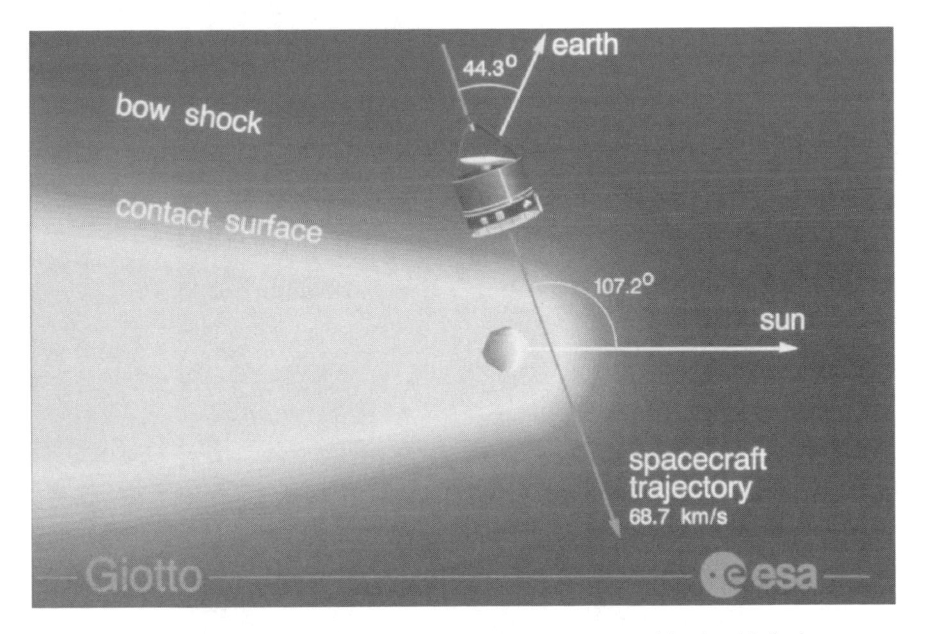

Figure 6.8. Schematic of *Giotto*'s encounter geometry with the Halley's comet. Illustration: European Space Agency.

In order to make the spacecraft stable and last as long as possible, and to exploit the design heritage of a previous ESA satellite named *GEOS*, the satellite was spin-stabilized at 15 rpm. The spacecraft was oriented to point its "Whipple Bumper" dust shield in its direction of flight relative to the comet. The bumper (named after comet pioneer Fred Whipple) was a two-layer shield, with a thin (1 mm) aluminium sheet mounted 25 cm ahead of a thicker Kevlar-reinforced layer some 12 mm thick. The idea was that a hypervelocity dust particle would tear through the forward sheet, being broken up and at least partly vaporized as it did so. The particle was thus turned into a shotgun blast of expanding vapor and smaller particles, which splashed harmlessly over a wide area of the rear layer.

Figure 6.9. *Giotto* spacecraft mounted on fixture for ground testing. The spacecraft's flight direction is downwards—note the flat plate dust shield at bottom, spaced 25 cm ahead of the thicker second layer. Some of the instruments are visible—the dust mass spectrometer inclined at the right, and the white cylindrical camera baffle is visible at the left. Above these is the cylindrical solar cell array for power, and above that the dish antenna. The antenna feed is mounted at the apex of the tripod. Photo: ESA.

Because the flight direction, along which the spin axis and the shield had to be aligned, was not aligned with Earth, the dish antenna was canted at an angle, and had to be despun. A precision-controlled motor spun the antenna backwards compared with the spacecraft, such that the radio beam stayed fixed in space, pointed at the Earth. The beam's pointing suffered, however, during the encounter.

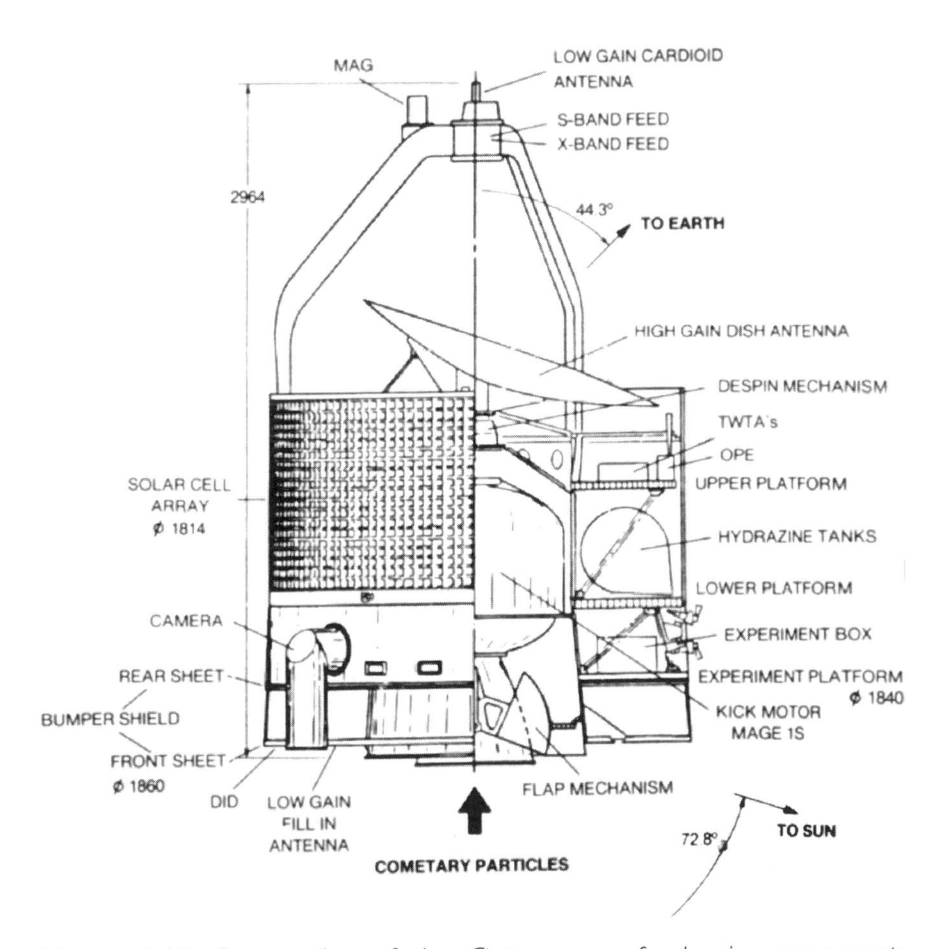

Figure 6.10. Cross-section of the *Giotto* spacecraft, showing components mounted on shelves around the central rocket motor (which was empty after the spacecraft departed Earth). Notice the sphere-cone tanks for hydrazine fuel for the spacecraft's thrusters—the spin forces the propellant outwards where it can be drawn through the propulsion system's plumbing. Illustration: European Space Agency.

The spacecraft had a camera—in fact the first CCD (charge coupled device) camera flown on a planetary mission; although such electronic imaging devices are very common today in digital cameras and phones, they were a novelty indeed in 1986. This Halley Multicolor Camera (HMC) looked outwards onto a mirror that allowed it to peer around the dust shield. The mirror, mounted in a white cylindrical baffle, could be rotated to allow the camera to track the comet nucleus as *Giotto* flew past the comet.

The attitude dynamics of *Giotto* during its encounter were measured using several different techniques. First, the satellite had a sun sensor which recorded the angle of the Sun from the spin axis by timing the crossing of the Sun through two slits, one along the spin axis and the other inclined. A star mapper worked in much the same way (Fertig et al., 1988).

Secondly, the strength of the received radio signal on the ground showed short-term variations which indicated changes in the position of the Earth in the antenna beam pattern (Bird et al., 1988). Additionally, the received radio signal varied in frequency due to the Doppler shift. Most of the shift was due to the predictable motion of the spacecraft relative to the Earth due to their different orbits around the Sun, but some short-term variations occurred if there was a wobble along the line of sight due to any coning motion, as well as changes in the orbital velocity due to drag caused by dust collisions.

The world watched as *Giotto* closed in. (I remember staying up to watch the event on television—it was around midnight on 13/14 March 1986. I was sixteen years old, and the event helped determine my career as an aerospace engineer and planetary scientist.) Hours before closest approach, the magnetometer and other instruments felt the effects of the comet on the plasma environment in space. The first dust impacts were detected about 70 minutes prior to closest approach, at a distance of 250,000 km (McDonnell et al., 1986).

The HMC began to feed pictures. A surprise was that far from being a "dirty snowball," the comet nucleus was as black as coal. The camera automatically tracked the brightest feature and as a result locked onto a bright jet of dust blasting out from the comet. The occasional tick of dust impacts rose to a crescendo of several hundred impacts per second.

The datalink was lost 7.6 s before closest approach—many people at the time thought the spacecraft had been smashed. But data began to be recovered 32 minutes later: *Giotto* had survived! The link was lost because the radio beam had been nudged off the Earth, and once the nutation dampers brought the attitude back under control, the link was regained. In fact, recordings of the radio signal made at the radio telescope (the 64 m dish in NASA's Deep Space Network at Tidbinbilla, Australia) showed that the transmission was always present, albeit at too low a level to permit recovery of the data. Several things appear to have happened, which took several months of detective work to untangle.

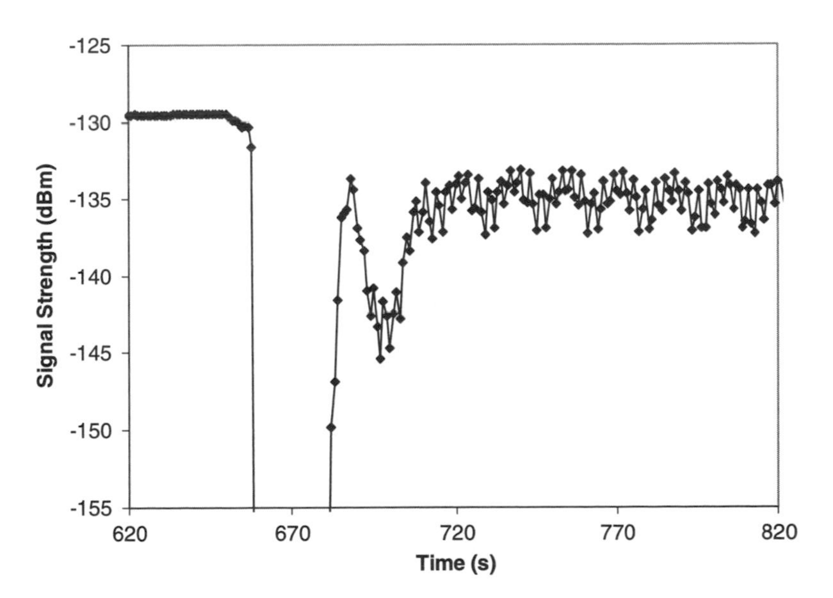

Figure 6.11. A plot of the received signal level (in dB referenced to 1 milliwatt) at the DSS-43 (Tidbinbilla) ground station. Modulation due to nutation is evident after encounter.

The antenna despin motor appears to have been desynchronized by some sort of electrical anomaly, perhaps an electrical discharge associated with the explosive formation of a small plasma cloud by a dust impact. The imperfect despin caused the Earth to drift out of the main beam of the antenna.

Another event was caused by an impulsive torque which misaligned the angular momentum vector with the spin axis, and thus caused a nutation. The spacecraft began a nutation with an angle of about 0.9 degrees and a period of 3.223 s.

The spin axis moment of inertia I_3 was 282 kgm^{-2}, while about the transverse axes I_T was 225–235 kgm^{-2}. The nutation period P_n and the spin period ($P_3 \sim 4$ s) are related by the nutation angle θ and moments of inertia as follows:

$$P_n = (I_T/I_3)\cos\theta\, P_3$$

The spacecraft was equipped with nutation dampers (fluid in loop) with a time constant of 2184 s; the recovery of the data link after about 2000 s makes perfect sense.

Exactly how large a dust particle caused the nutation isn't known—the angular momentum impulse for a given dust impact will depend on where the particle hit. Because of the telemetry loss around closest approach, there was no information from the dust impact detector itself. If the impact was dead center, there would be no angular momentum increment, while an impact at the edge would have the greatest possible angular impulse. For a given angular momentum impulse (inferred to be 7.8 kgm^{2}s^{-1}), the mass of particle required is smaller the further from the edge it hit. Taking into account the distribution of dust particle sizes measured by an impact detector on the front shield, the most probable combination of parameters for the observed change in angular momentum is a particle between 110 and 150 mg impacting in the outer 25 cm of the shield.

A dust impact on the front shield cannot change the spin rate, however, since the angular momentum increment is always orthogonal to the spin angular momentum vector. The change in spin period from 3.998 s to 4.0009 s therefore requires a different explanation. This was named the "HMC windmill effect."

When a dust particle hits the inclined camera baffle, the explosion of vaporized material expands away from the surface, as if the particle were diffusely reflected. The reaction force of this "rocket" or explosion effect therefore has a net direction along the normal to the surface of the baffle at that point. Since the baffle was inclined, a component of this reaction force was therefore in the spin plane, and thus could provide a torque to change the spin rate.

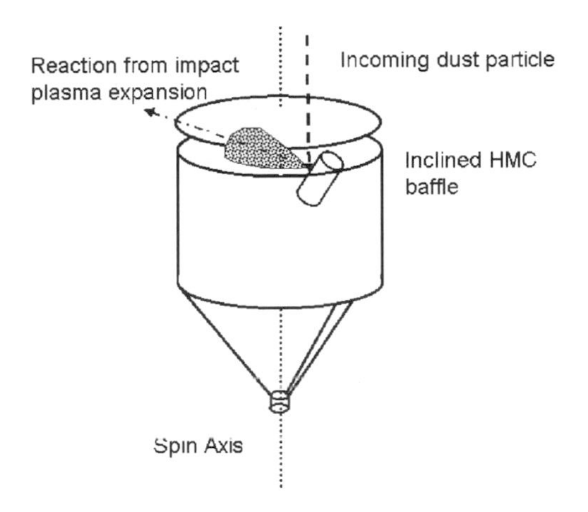

Figure 6.12. HMC windmill effect. A schematic of how the (shaded) explosion cloud from a dust impact on the inclined HMC baffle could provide a torque around the spin axis to change the *Giotto* spin period.

Notice that as well as providing an impulse component orthogonal to the direction of the incoming particle, this explosion also adds to the impulse along the velocity vector of the particle: a dust grain impacting normally on a surface will deposit its own momentum, plus an extra impulse due to the expansion of the vapor back along the direction the grain came from. Space dust experts denote this additional impulse by a "momentum enhancement factor" e, which will depend on the impact velocity and on the target material and how easily it is thrown off the surface.

The total mass of dust encountered by the spacecraft is estimated to be something under 1.93 g, which caused the 573 kg spacecraft to slow down by 23.05 s. Strictly speaking, 1.93 g = $(1 + e)$ times the actual mass, and e was of the order of 10–40. This should have caused a

Doppler shift of 4.64 Hz in the spacecraft's 8.428 GHz X-band signal—although high precision measurements using two-way transmissions from ground before and after the encounter confirm this, the real-time Doppler measurements using the spacecraft's on-board oscillator actually show a 16.9 Hz shift—apparently stresses on the oscillator during the violent encounter caused its frequency to drift.

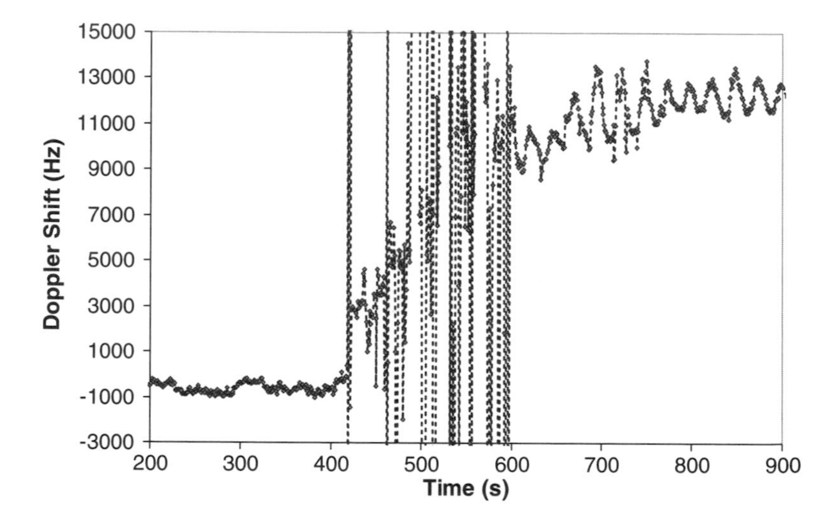

Figure 6.13. A plot of the Doppler shift on *Giotto*'s noncoherent X-band downlink during the Halley encounter. Since the spacecraft was flying away from the Earth, the dust collisions which slowed it down caused the speed of recession to decrease, thus "blue-shifting" the radio link, or increasing its frequency. The observed shift is higher than expected, due to impact-induced drift on the radio link's oscillator. Notice the spikes due to signal dropouts in the thick of the encounter, and the modulation of the frequency after encounter due to nutation.

The post-encounter dynamics also indicated a persistent wobble after the nutation had damped down. This wobble requires a change in the mass distribution of the satellite, such that the nominal spin axis (to which the sun sensor data referred) was no longer the axis of maximum moment of inertia. Some numerical experiments with how the sun sensor data should look under different scenarios suggested that the camera baffle had been torn off! This interpretation was also supported by study of the solar array currents—previously the currents dropped slightly during a spin period when part of the array was shadowed by the baffle; this shadowing no longer occurred after encounter. And of course the camera was no longer working.

Other Encounters

The two *VEGA* spacecraft, which were not spin-stabilized or equipped with bumper shields, suffered considerable damage, despite being rather further away (~6000 km) during their closest approaches a couple of days before. As well as losing many scientific instruments, they lost some 40% of their solar array capacity.

One of the two spin-stabilized Japanese spacecraft, *Suisei*, felt bumps from the comet. This 140 kg satellite never approached closer than 150,000 km, yet suffered 2 dust impacts which were detected by impulsive changes in spin dynamics. The satellite was not in contact with Japan's Usuda ground station at the time—the changes were observed in sun sensor data which had been recorded in on-board memory.

Figure 6.14. The Japanese *Suisei* spacecraft. Like *Giotto*, it was equipped with a despun high-gain antenna and a cylindrical solar array. Photo: ISAS/JAXA.

One impact 12 minutes before closest approach changed the spin period from 9.184 s to 9.157 s and shifted the spin axis by about 10 degrees. (Since *Suisei* was not expected to encounter dust, it had no shield, and its spin axis was orthogonal to the Sun.) The second particle a half hour later changed the spin period by 0.001 s. Both impacts excited a nutation which an on-board damper caused to decay with a timescale of ~20 minutes. Given the relative speed of 73 km/s, the moment of inertia of the satellite at 31 kgm^2 and the radius of the satellite at 0.7 m, the dust particles were estimated to have masses of a few milligrams and a few tens of micrograms, respectively. The uncertainties derive from ignorance of where on the satellite the particles hit, and on the momentum enhancement factor. In the case of the smaller impact, the inferred impulse orthogonal to the velocity vector of the dust grains suggests a very large momentum enhancement factor (>300), suggesting that perhaps the dust impact blew off a solar cell.

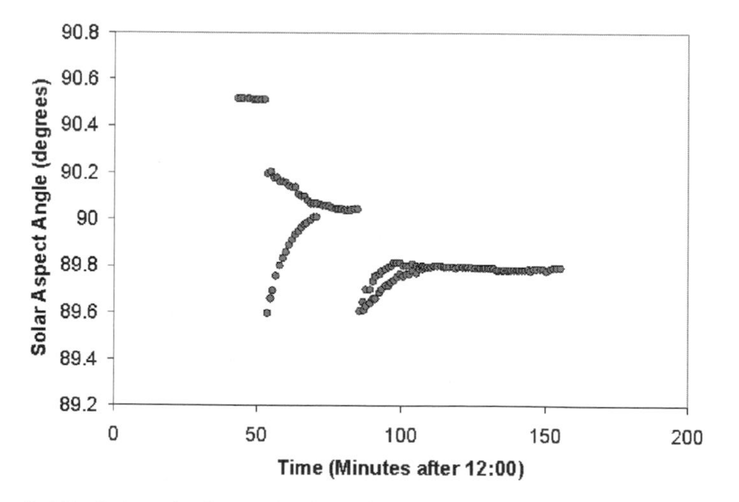

Figure 6.15. Spin axis determinations from the *Suisei* sun sensor. The impulsive changes in spin axis, and the nutation (shown as a bifurcation in the solar aspect angle measurements) which damps down thereafter, were caused by two dust impacts, the second rather smaller than the first. (Uesugi, 1986)

There is a short postscript to the Halley story. *Giotto* was placed into hibernation: although some systems were damaged, the spacecraft was basically functional. When it flew by the Earth in 1990 it was redirected using the Earth's gravity to an encounter with another comet, P/Grigg–Skjellerup. This encounter, in the summer of 1992, was much less violent, with only a couple of dust impacts being detected on the front shield. The encounter geometry this time exposed the cylindrical solar array to the dust velocity (14 km/s), and during the encounter the radio science investigation, again using signal strength and Doppler variations (Patzold et al., 1993), found an impulsive change in spin rate (the spin period dropped by 0.8 ms) and the onset of a nutation of 0.1 degrees. These suggest that a relatively large dust particle, of 20–39 mg, impacted the spacecraft either on the cylindrical part or on the tripod holding the magnetometer and the antenna feed.

References

Belton, M. J. S., W. H. Julian, A. Jay Anderson, B. E. A. Mueller, The spin state and homogeneity of comet Halley's nucleus, *Icarus* 93, 183–193, 1986.

Belton, M. J. S., N. H. Samarasinha, Y. R. Fernández, and K. J. Meech, The excited spin state of comet 2P/Encke. *Icarus* 175, 181–193, 2005.

Bird, M. K., M. Patzold, H. Volland, P. Edenhofer, H. Buschert, and H. Porsche, *Giotto* spacecraft dynamics during the encounter with Comet Halley, *ESA Journal* 12, 149–169, 1988.

Chesley, Steven R., Steven J. Ostro, David, Vokrouhlický, David, Čapek, Jon D., Giorgini, Michael C., Nolan, Jean-Luc, Margot, Alice A., Hine, Lance A. M., Benner, and Alan B., Chamberlin, Direct detection of the Yarkovsky effect via radar ranging to asteroid 6489 Golevka, *Science* 302, 1739–1742, 2003.

Fertig, J., X. Marc, and J. Schoenmaaekers, Analysis of *Giotto* encounter dynamics and post-encounter status based on AOCS data, *ESA Journal* 12, 171–188, 1988.

Hudson, R. S., and S. J. Ostro, Shape and non-principal-axis spin state of asteroid 4179 Toutatis. *Science* 270, 84–86, 1995.

Hudson, R. S., and S. J. Ostro, Photometric properties of asteroid 4179 Toutatis from lightcurves and a radar-derived physical model. *Icarus* 135, 451–457, 1998.

Lorenz, R. D., and J. N. Spitale, The Yarkovsky effect as a heat engine, *Icarus* 170, 229–233, 2004.

McDonnell, J. A. M. et al., Dust density and mass distribution near comet Halley from *Giotto* observations, *Nature* 321, 338–341, 1986.

Mueller, B. E. A., N. H. Samarasinha, and M. J. S. Belton, The diagnosis of complex rotation in the lightcurve of 4179 Toutatis and potential applications to other asteroids and bare cometary nuclei, *Icarus* 158, 305–311, 2002.

Ostro, S. J., R. S. Hudson, K. D. Rosema, J. D. Giorgini, R. F. Jurgens, D. K. Yeomans, P. W. Chodas, R. Winkler, R. Rose, D. Choate, R. A. Cormier, D. Kelley, R. Littlefair, L. A. M. Benner, M. L. Thomas, and M. A. Slade, Asteroid 4179 Toutatis: 1996 radar observations, *Icarus* 137, 122–139, 1999.

Patzold, M., M. K. Bird, and P. Edenhofer, The change of *Giotto*'s dynamical state during the P/Grigg-Skjellerup flyby caused by dust particle impacts, *J. Geophys. Res.* 98, 20,911–20,920, 1993b.

Pravec, P., and A. W. Harris, 2000 fast and slow rotation of asteroids. *Icarus* 148, 12–20, 2000.

Rubincam, David P., Radiative spin-up and spin-down of small asteroids, *Icarus* 148, 2–11, 2000.

Slivan, S., Spin vector alignment of Koronis family asteroids, *Nature* 419, 49–52, 2002.

Spencer, J. R., L. A. Akimov, C. Angeli, P. Angelini, M. A. Barucci, P. Birch, C. Blanco, M. W. Buie, A. Caruso, V. G. Chiornij, F. Colas, P. Dentchev, N. I. Dorokhov, M. C. De Sanctis, E. Dotto, O. B. Ezhkova, M. Fulchignoni, S. Green, A. W. Harris, E. S. Howell, T. Hudecek, A. V. Kalashnikov, V. V. Kobelev, Z. B. Korobova, N. I. Koshkin, V. P. Kozhevnikov, Y. N. Krugly, D. Lazzaro, J. Lecacheux, J. MacConnell, S. Y. Mel'nikov, T. Michalowski, B. E. A. Mueller, T. Nakamura, C. Neese, M. C. Nolan, W. Osborn, P. Pravec, D. Riccioli, V. S. Shevchenko, V. G. Shevchenko, D. J. Tholen, F. P. Velichko, C. Venditti, R. Venditti, W. Wisniewski, J. Young, and B. Zellner. The lightcurve of 4179 Toutatis: Evidence for complex rotation. *Icarus* 117, 71–89, 1995. (The collaborative nature of asteroid photometry projects with many individuals contributing a few datapoints to create a coherent picture of the rotation state is evident in the length of the author lists of some of these papers!)

Uesugi, K., Collision of large dust particles with *Suisei* spacecraft, In *Proceedings of the 20th ESLAB Symposium on the Exploration of Halley's Comet.* Volume 2: Dust and Nucleus p. 219–222. European Space Agency, 1986.

7

Planetary Probes and Spinning Parachutes

Many vehicles sent into the atmospheres of other planets (or indeed back into the Earth's atmosphere from above) have used spin for stabilization.

The kinetic energy of an object in orbit or on an interplanetary trajectory (with a velocity of perhaps 7 km/s—Mach 25—in the first case, or up to 50 km/s in the second) is formidable. Even in the former case, this energy is comparable with or exceeds the latent heat of evaporation of most materials. Thus in order for the probe not to melt or evaporate, most of this energy must be dissipated somewhere other than the spacecraft.

The approach usually used is to make the probe or at least its heat shield a blunt body, with a large radius of curvature. This blunt shape causes a strong shock wave in the hypersonic flow; the relative airflow is decelerated (leading to conversion into heat) in this shockwave, rather than at the heat shield itself. While some fraction of the kinetic energy is still transferred to the body by convection and radiation (the shock layer will often glow brightly or even brilliantly—this is after all the process that makes meteors shine), this fraction of typically a few percent is much more manageable. Were the entry body to be sharp-nosed, it would be more stable aerodynamically, but the heat loads on the nose would be unbearably intense.

Because heat shielding is generally heavy, it is desirable to only need protection on one side of the vehicle. (Some shielding is still required on the back side, to protect against recirculating flow and in particular against radiation from the hot wake or "meteor trail".) This in turn requires the orientation at entry to be controlled (so-called 3-axis stabilization) by active attitude control with thrusters (as done with the Apollo capsules and the Viking landers on Mars), by passive weathercock stability, or by spin-stabilization. Obviously the latter option, one of the most popular, comes under the purview of this book.

Some Russian entry probes to the planet Venus had a spherical or egg shape, with heat shield material all around. The center of mass was offset from the center, giving a preferred orientation. A more obvious example of this type of stability was NASA's two *Deep Space 2* (*DS-2*) *Mars Microprobes*, a pair of tiny (4 kg, 30 cm) entry shells designed to hit the ground at some 200 m/s and bury penetrators into the ground in 1999. These lightweight entry shells had their penetrator payloads placed as far forward as possible, giving the system a very strong weathercock stability. In fact, the penetrator nose was made of tungsten (an expensive metal with the remarkably high density of 19 times that of water!) for the express purpose of moving the center of mass as far forward as possible for aerodynamic stability.

Figure 7.1. The *Mars Exploration Rover* inside its entry shield (same design as the *Pathfinder* entry shield) and the carrier spacecraft. Notice how the spacecraft dynamically resembles a flat disk. This photo (NASA PIA04823) was taken around Halloween 2002—note the decoration on the fuel tank.

Among spin-stabilized entry probes have been the similar *Pathfinder* and *Mars Exploration Rover* missions. During their cruise through space they are dynamically stabilized by having a moment-of-inertia ratio of 1.27. Interestingly, the sensitive accelerometers on board recorded a slight periodic signature before *Pathfinder* encountered the atmosphere of 20 or 30 micro-g amplitude (e.g., Spencer et al., 1999). The period of the signal was 110 s—comparing this with the 30 s rotation period and the inertia ratio confirms that this is a slight nutation signal: $\omega_{obs} = (1 - I_{zz}/I_{xx})\omega_{spin}$.

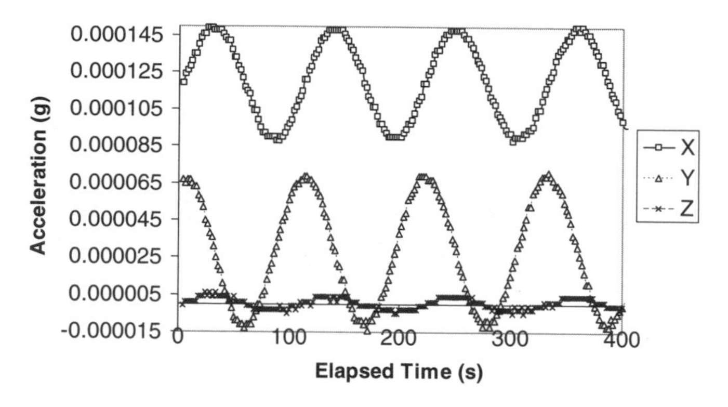

Figure 7.2. Pre-entry accelerations recorded by the accelerometers on the *Mars Pathfinder* spacecraft, indicating a slight nutation.

Ablation of Entry Vehicles

Intuitively one might expect that the spin of an axisymmetric (no fins) vehicle entering an atmosphere from space should not be substantially affected. One might imagine a small spin-down due to skin friction torques, but even these should be modest.

However, where very large heat fluxes are experienced, such as the very energetic entry into the atmosphere of Jupiter, or where the probe's mass:area ratio is large in other atmospheres, it may happen that an ablative heat shield is needed, where a significant amount of material is "burned off" (more usually melted or sublimated).

A subtle effect is that of the heat shield construction. Some types of heat shield, such as the carbon-carbon type able to resist high temperatures, are made from tapes or cloths of carbon fiber material, sintered or baked together. If this material is wound onto the shield as a tape, this tape often forms a spiral pattern. If the spiral, or indeed the free end of the tape, becomes exposed during the entry, these odd surfaces can act to drive spin torques.

A dedicated rocket test (Kryvoruka and Bramlette, 1977) was made to investigate this effect. This sharp entry vehicle (7 degrees, ~1 m in length—characteristic of a warhead on a ballistic missile, rather than a planetary entry vehicle) was accelerated to Mach 8.8 on a 3-stage rocket. After separation at nearly 100 revolutions per second, the spin decayed to about 50 revolutions per second in only 40 seconds. This rapid spin-down is believed to be due to fluted patterns formed in the ablative heatshield. At one point in the flight (at T ~ 18 s, 5 s after separation) the other motions reached high rates of pitch and yaw, presumably due to passage through the roll–yaw resonance condition. Note that it is likely that this work was not inspired by planetary science needs, but rather because roll reversal of entry vehicles may be a source of delivery error in MIRV (Multiple Independent Reentry Vehicles) warheads from ballistic missiles. The effectiveness of a nuclear strike on hardened targets like missile silos is strongly dependent on how close the warhead can be delivered; unfortunately soft targets like cities are much more forgiving in this regard.

Tests in a hypersonic wind tunnel (McDevitt, 1971) had explored the formation of fluted patterns. A NASA Ames tunnel was operated at Mach numbers of up to 10, with the airflow heated by passing over a hot pebble bed before being expanded in a nozzle. The models, also slender entry vehicles, were mounted on a special air bearing in order to permit the vehicle to spin up or down. Because the speeds and durations required to reproduce the conditions of entry are difficult to reproduce, the ablation effects were explored by making the test vehicle from a readily ablatable material. Some tests were made of camphor (the material used to make mothballs); others used ammonium chloride, a salt that readily sublimes. Spin rates of the model were monitored initially by a magnetic tachometer, then later with a photodiode and a reflective surface on the model (it was noted that the magnetic tachometer in fact produced a braking torque due to eddy current damping—see chapter 5).

In a minute or so of operation, the ~530 g models would lose around 40 g of mass by ablation. This mass loss was often in the form of striated patterns—longitudinal grooves, cross-hatching and turbulent wedges. The models developed spin in either direction, without regard to the direction of any initial spin. Slightly different behavior was seen with models made from a material named Korotherm, which melts before it sublimes away. With this material, perhaps because of melt flow or perhaps because of the high coefficient of thermal expansion of the material, the models tended to spin up in the direction of any initial spin. Clearly the spin of ablating materials in hypersonic flow is a complex topic.

Spin of Parachute-Borne Planetary Probes

After a planetary probe enters an atmosphere and slows to conventional aerodynamic speeds, it is often desirable to retard its descent with a parachute, either to provide for a softer landing, or to prolong the time at high altitude for scientific measurements. But parachutes can cause spin if not symmetrically rigged. Thus, since the spin of a parachute is rather difficult to predict, it is often expedient to decouple it from the store with a swivel.

In cases where there is no such swivel, what can often happen is that the differential torques on the payload and on the chute (rather, the different angular accelerations—strictly speaking the torques could be the same, but since parachute and store have different moments of inertia, one would spin up faster than the other!) lead to a relative rotation. This tends to wind up the lines to the parachute, leading to some elastic storage of energy (the lines and chute act as a torsional spring). Shortly thereafter, either due to some transient drop in torque, or inertial overshoot, the torsion in the lines will arrest and reverse the relative rotation, such that the payload begins spinning the other way. This cycle of wind up can repeat many times. Similar effects can occur for suspended payloads such as astronomical telescopes under high-altitude balloons.

Venus Probes

The planet Venus has been explored by a series of probes developed in Russia and launched in the 1970s and 1980s, as well as one multiprobe mission by the U.S. in 1979. Generally these used a drag disk to modestly slow and stabilize their descent, which would be too slow in the thick 90-bar Venus atmosphere with a full-scale parachute. (Parachute materials would rapidly deteriorate in the hot lower atmosphere, where temperatures exceed the melting point of lead!) Where parachutes have been used, they are discarded early in the descent, before the hottest part of the atmosphere has been reached.

This approach was adopted by the Russian Venera probes. These were kept stable during their free-fall by a drag plate near the top. This sharp-edged disk causes the flow to separate at a constant location, whereas a smooth body might experience periodic vortex shedding, exciting undesirable motions.

Figure 7.3. *Venera 9* lander. The main equipment is in the spherical pressure vessel; the cylinder at top is a helical antenna. Note the crush ring for impact attenuation and the drag disk to retard and stabilize the descent. Note also the spiral antenna: the spiral configuration controls the radiation pattern from the antenna and the polarization of the signal. Image: NASA NSSDC.

The optical instruments on the U.S. Pioneer Venus probes required spin to sample the light flux in various directions to understand the scattering properties of the clouds, so a set of spin vanes were installed on the probes. These were a large probe (which did use a parachute briefly) and three small probes (which did not use parachutes at all), named after their target areas "Day," "Night," and "North." The "Night" probe's thermal flux sensors apparently recorded at least briefly indications that the spin axis was inclined, and the spin rate was 7.5 rpm (Suomi et al., 1979). The very thick clouds on Venus meant that direct sunlight was quickly extinguished; optical sensors provided only very brief indications of spin before the large probe sank beneath the clouds (M. Tomasko, personal communication). In any case, the scientific requirements on Pioneer Venus's spin were not especially harsh, nor was the spin documented in any detail.

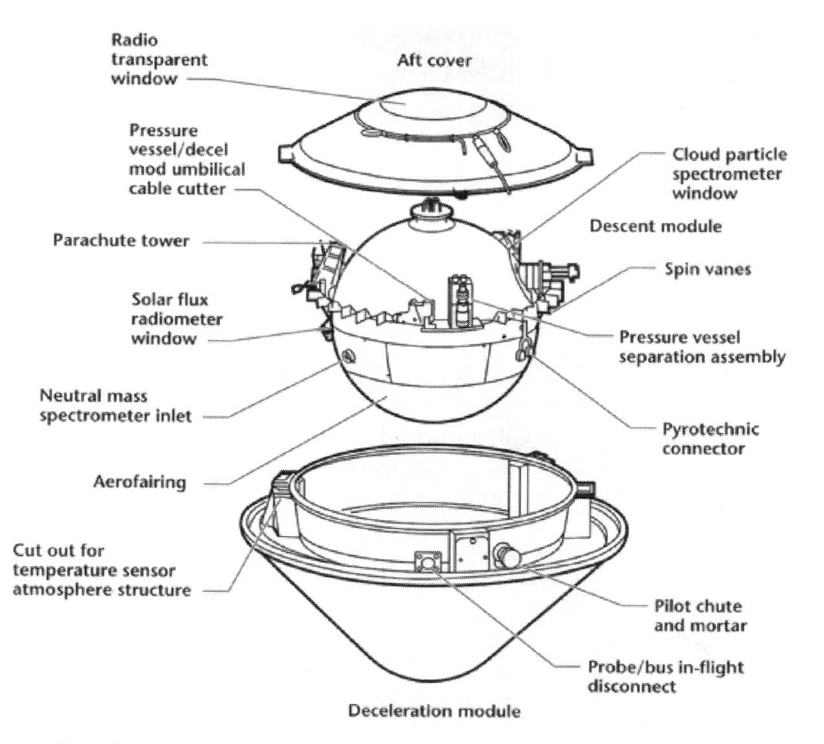

Figure 7.4. *Pioneer Venus* large probe. The probe was built as a rigid sphere to resist the hot high-pressure atmosphere. After the heat shield ("deceleration module") has fallen away, the spin vanes cause a gentle rotation. Image: NASA.

The Pioneer Venus probes did neatly exploit spin from their carrier spacecraft. The dispersed aimpoints at different locations were achieved in part by releasing the probes from the carrier at precisely controlled times, such that the probes were slung off with a desired sideways component of velocity.

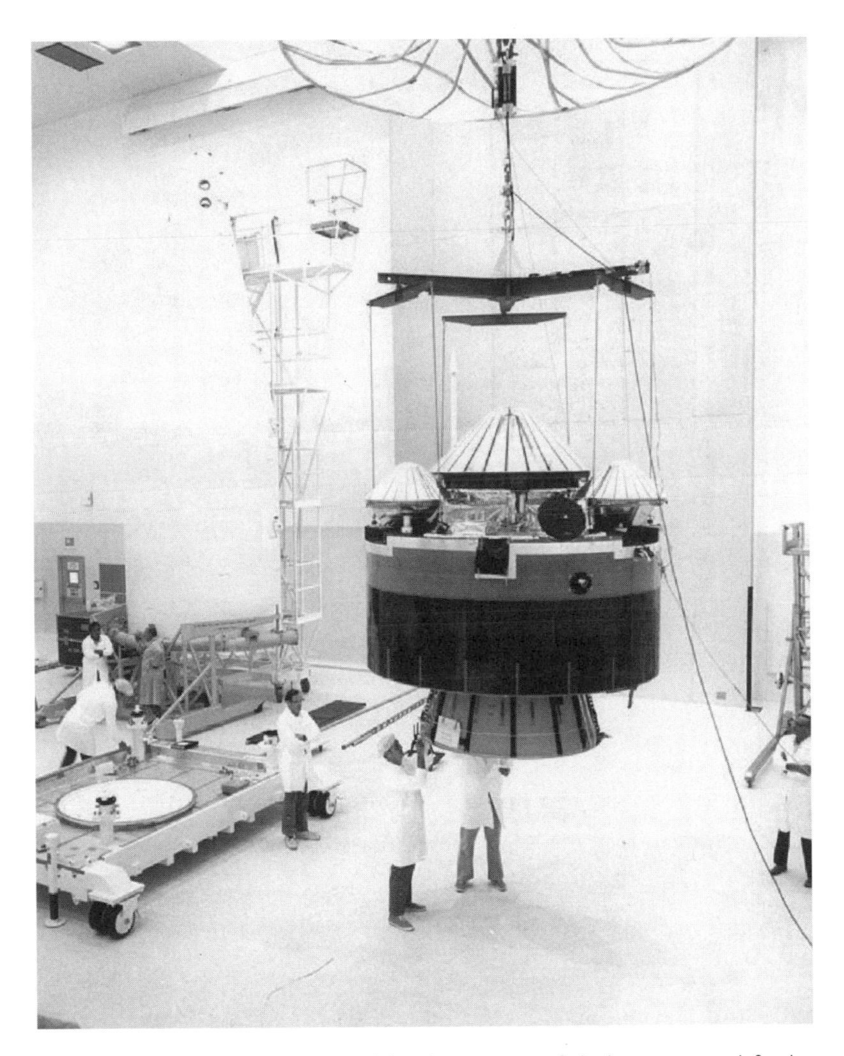

Figure 7.5. The *Pioneer Venus* multiprobe spacecraft being prepared for launch at Kennedy Space Center. The "bus" spacecraft spun; the large probe was simply pushed out along the spin axis of the spacecraft, while the small probes (of which two are visible) were released to fly out sideways using the rotation of the bus to achieve the desired dispersion. NASA photo KSC-78P-0171.

Figure 7.6. Illustration of the PV multiprobe releasing the large probe (top right) and later the three small probes. The lateral release (note the curved arm to the right of the carrier spacecraft) provided the desired aimpoint distribution. NASA image AC78-9245 (artwork by Paul Hudson) courtesy NASA Ames Research Center.

Galileo

The *Galileo* probe was equipped with three spin vanes to guide its spin rate during parachute descent. A peculiarity of the *Galileo* probe compared with Venus, Titan, and Mars missions is that the entry speed into Jupiter's atmosphere was very high—some 50 km/s (a factor of 7 more than the other missions). Aerodynamic heating scales as the cube of velocity, and the high velocity caused by Jupiter's enormous gravity means the heating problem is especially acute for Jupiter probes; indeed, the probe had to be aimed at low latitudes on the dusk side of Jupiter to keep the velocity manageably low. (As the probe falls in from infinity, it has a speed of 60 km/s; the large planet Jupiter rotates fairly quickly, such that the circumferential speed is some 10 km/s. By choosing the dusk limb, the speed of the probe relative to the rotating air is reduced to 50 km/s. Were the morning limb aimed at, the speed would be 70 km/s, and the heat fluxes nearly 3 times higher than at the receding limb.)

A result of the high heat loads is that *Galileo* needed a formidable heat shield, taking up around half of the probe's mass. Some 89 kg of this heat shield was burnt away during entry. The possibility exists when such large mass loss is encountered that it may not be uniform, but flutes or grooves might be formed in the heat shield material, causing the probe to quickly spin up in one direction or another.

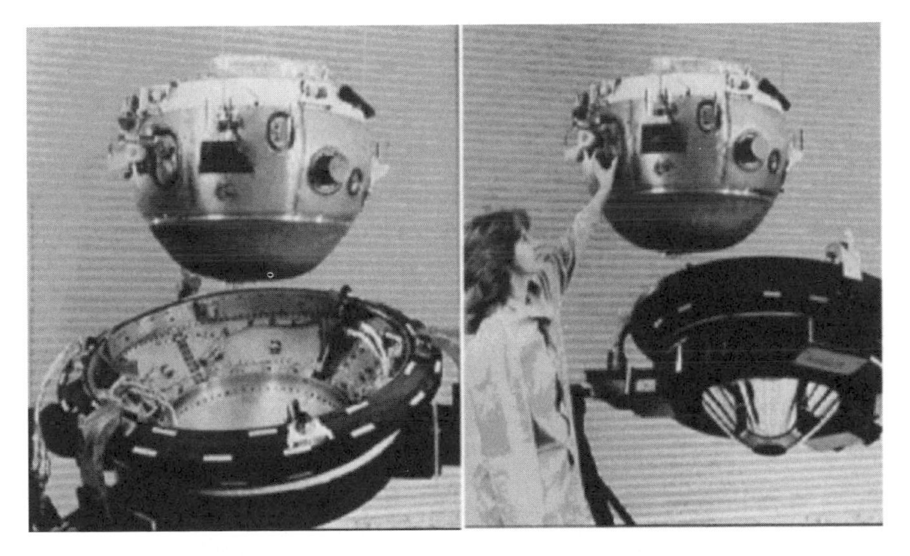

Figure 7.7. *Galileo* probe being prepared for launch. The heat shield is below. One of the spin vanes is visible on the hemispherical descent module, just right of center. NASA photos S89-45784 and S89-45785, courtesy NASA JSC.

The probe was released in space with a spin of 10.4 rpm, but after entry the probe was found to be spinning at some 33.5 rpm (Lanzerotti et al., 1998). This was determined using the lightning and radio emissions detector (LRD). This was able to measure (inside the heat shield, which obscured optical instruments) the spin-modulation of Jupiter's magnetic field.

The probe's spin design was principally to ensure a spin of less than 50 rpm, to avoid excessive Doppler shift on the radio link: scientific measurements would be somewhat degraded if the spin fell outside the range 0.25–40 rpm. To achieve this, at least after the uncertain transient conditions at deployment, the probe was equipped with three spin vanes, two mounted at about 13 degrees, and one at about 3 degrees (the orientations being checked with a mirror and theodolite). The three vanes were also used to balance asymmetric aerodynamic loads from various protrusions from the probe and thus did not have quite the same orientation.

Parachute deployment did not appreciably slow the spin (the parachute spin was decoupled from the probe by a swivel), but the spin drops quickly when the heatshield was released and the spin vanes exposed.

Although the spin modulation of the magnetic field does not indicate the sense of the rotation, it can be assumed that this was in the sense demanded by the spin vanes (clockwise, looking down). There was no indication in the spin data of a zero spin period.

In fact the dynamics of the *Galileo* probe have not been analyzed in a systematic way, by fusing data from different sensors. One indication from optical sensors is that the probe could have been tilted by some 19 degrees (Sromovsky et al., 1998) early in the descent above the clouds when spin was around 30 rpm. On the other hand, a modulation of the signal strength of the probe radio signal varying from a 14 s period to 50 s has (perhaps incorrectly) been attributed to probe spin, with a shorter period variation attributed to swing under the parachute. Cleary substantial uncertainties, if not discrepancies, remain.

Another spin-related observation from *Galileo* merits comment. The violent deceleration during entry was recorded by an accelerometer in order to measure the density structure. During the parachute descent, the accelerometer was switched into a more sensitive mode in order to measure atmospheric turbulence. The axial (i.e., nominally vertical) accelerometer measured a local gravitational acceleration slightly lower than would be expected knowing Jupiter's radius, mass and bulk

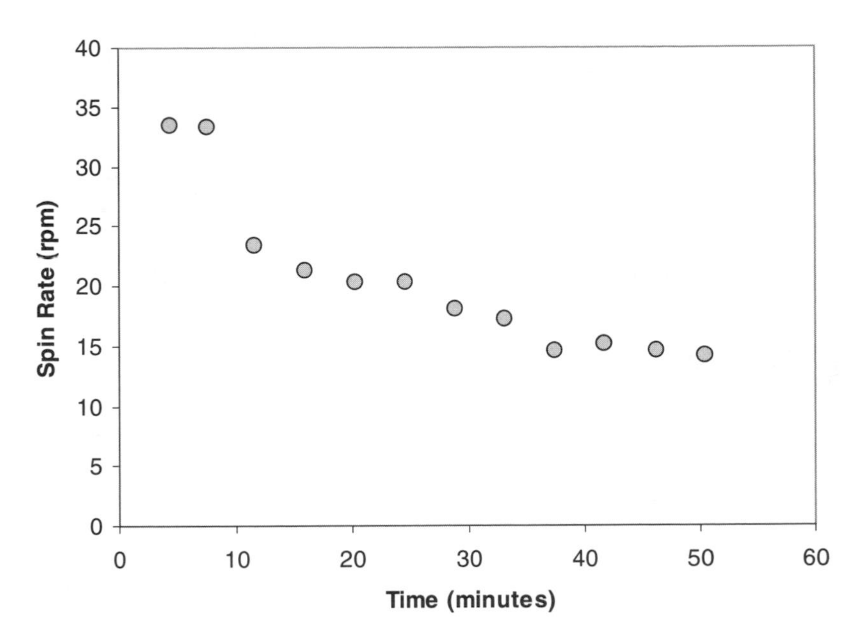

Figure 7.8. Spin rate of the *Galileo* probe as determined by its lightning and radio emissions detectors. The sharp drop at ~10 minutes corresponds to the release of the (perhaps fluted) heat shield and the exposure of the spin vanes.

rotation rate (Seiff et al., 1997). The difference was due to the centripetal acceleration due to the surprisingly strong deep zonal winds—in effect the winds swept the probe around as if the local rotation rate were higher than that of the planet as a whole.

Huygens

The *Huygens* probe to Titan, like many others, was spin stabilized during its dormant coast through space. During its 2.5-hour parachute descent, its spin about a vertical axis was also controlled in order to pan around the field-of-view of its side-looking camera.

The expected spin rate profile is complicated. In a sense, the installation of spin vanes along the equator of the probe yields a "demanded" spin rate which depends on the orientation of the vanes and on the descent speed of the probe. A given spin rate will produce an angle of

Figure 7.9. Artist's impression of the *Huygens* probe descending under its parachute in Titan's atmosphere. By Mark Robertson-Tessi and Ralph Lorenz. See http://www.lpl.arizona.edu/~rlorenz.

attack on the vanes—the probe will tend to be spun up or down to where that angle of attack leads to zero net side-force (spin torque) on the vanes. In other words, the vane setting angle determines a spiral which the probe, in the absence of other torques, would eventually tend to follow.

It is assumed that the rotation of the probe itself is decoupled from that of the parachute; the riser incorporates a swivel which transmits a negligible torque. However, the probe has a substantial moment of interia, and thus takes some time to respond to changes in the demanded speed.

This effect can be seen in a drop test ("SM2: Special Model 2"—Jakel et al, 1996) that was conducted to verify the deployment sequence of *Huygens* parachutes. In order to match the Titan deployment conditions in terms of Mach and Reynolds number as closely as possible, this test was done from a stratospheric balloon. (Since the parachute test, the *SM2* unit has been used frequently for displays at airshows and conferences, etc.)

Color Plate

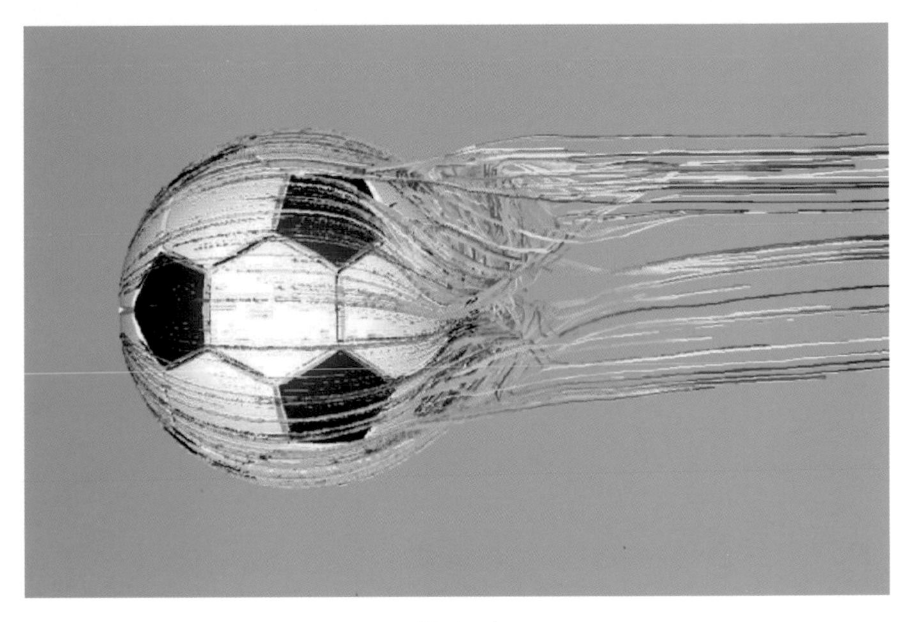

Plate 1

Plate 2

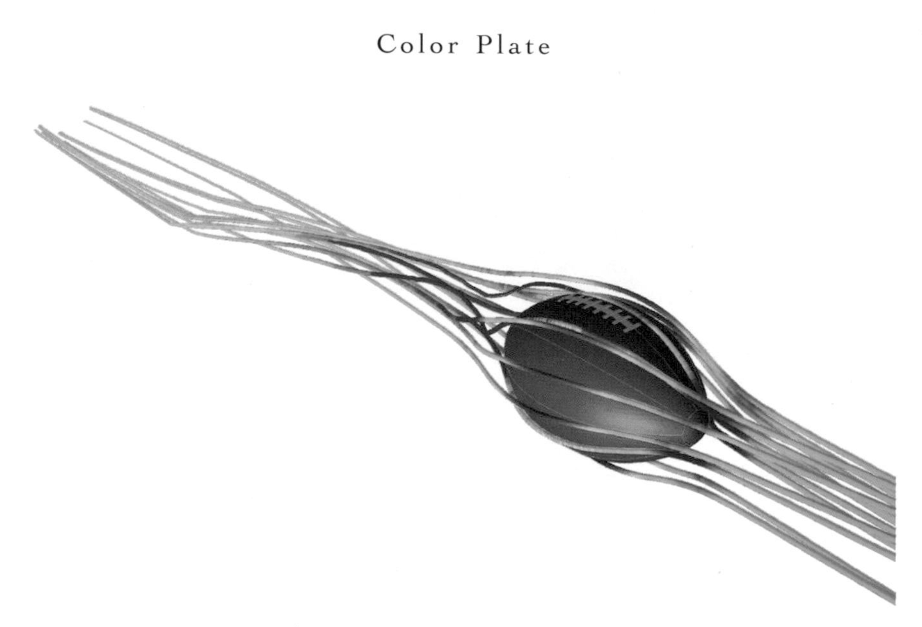

Plate 3

Plate 4

Plate 5

Plate 6

Plate 7

Plate 8

Plate 9

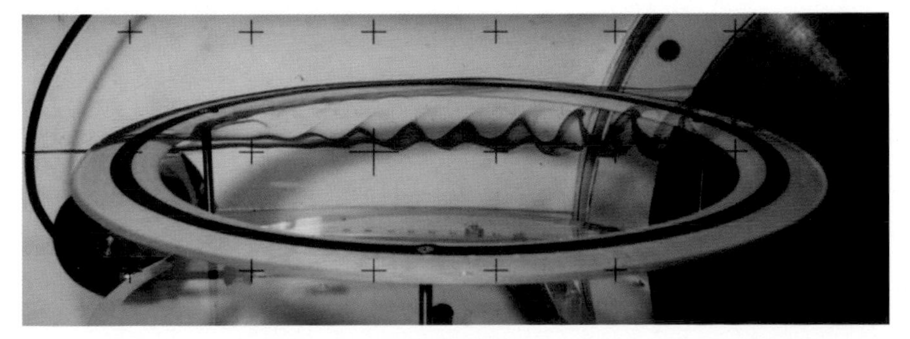

Plate 10

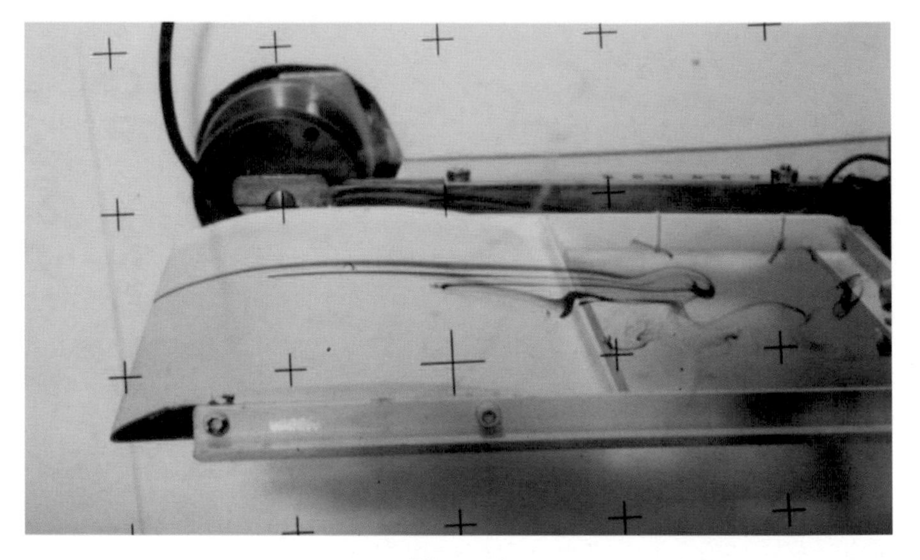

Plate 11

Plate 12

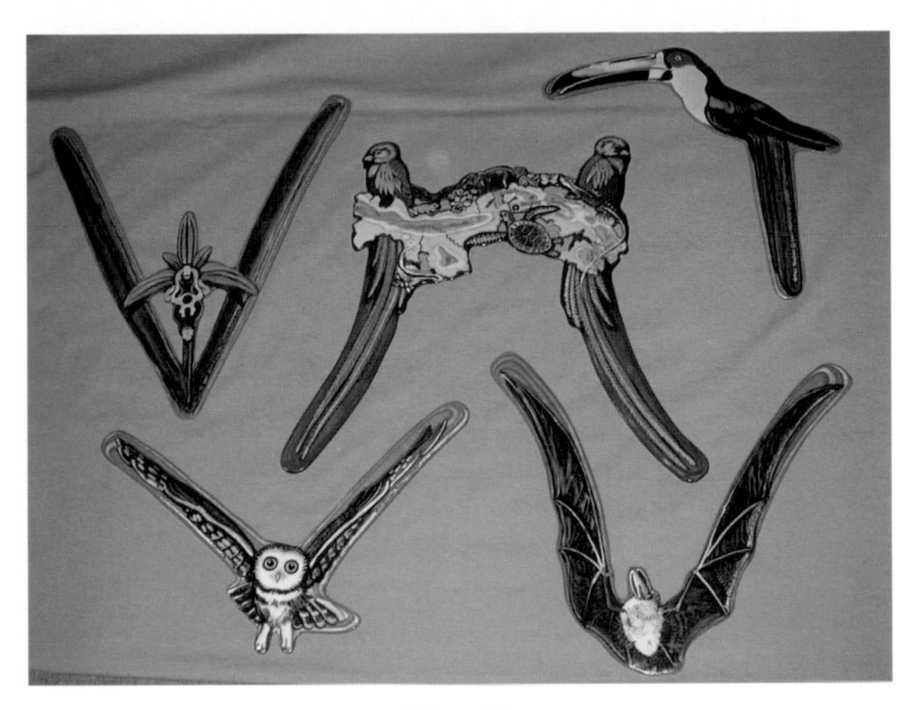

Plate 13

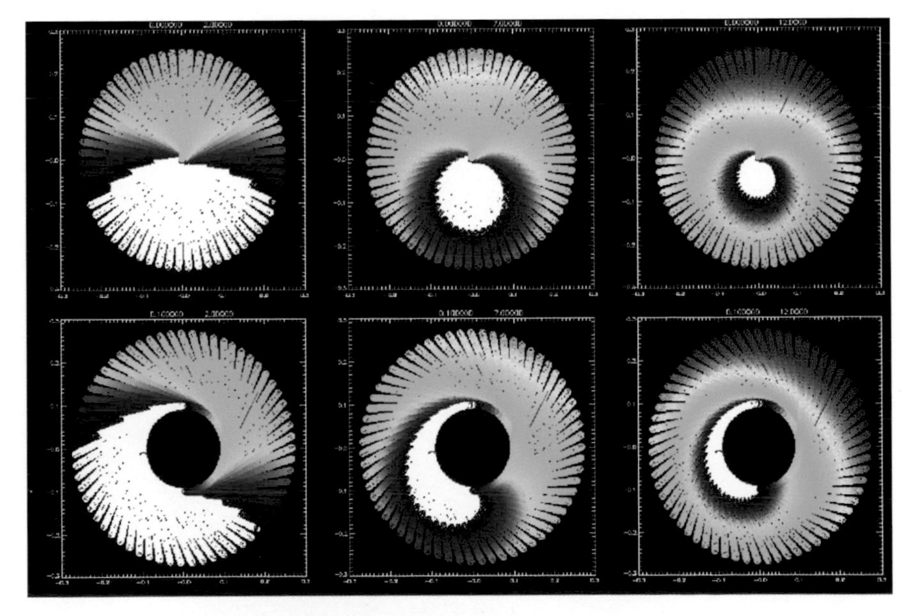

Plate 14

Plate 15

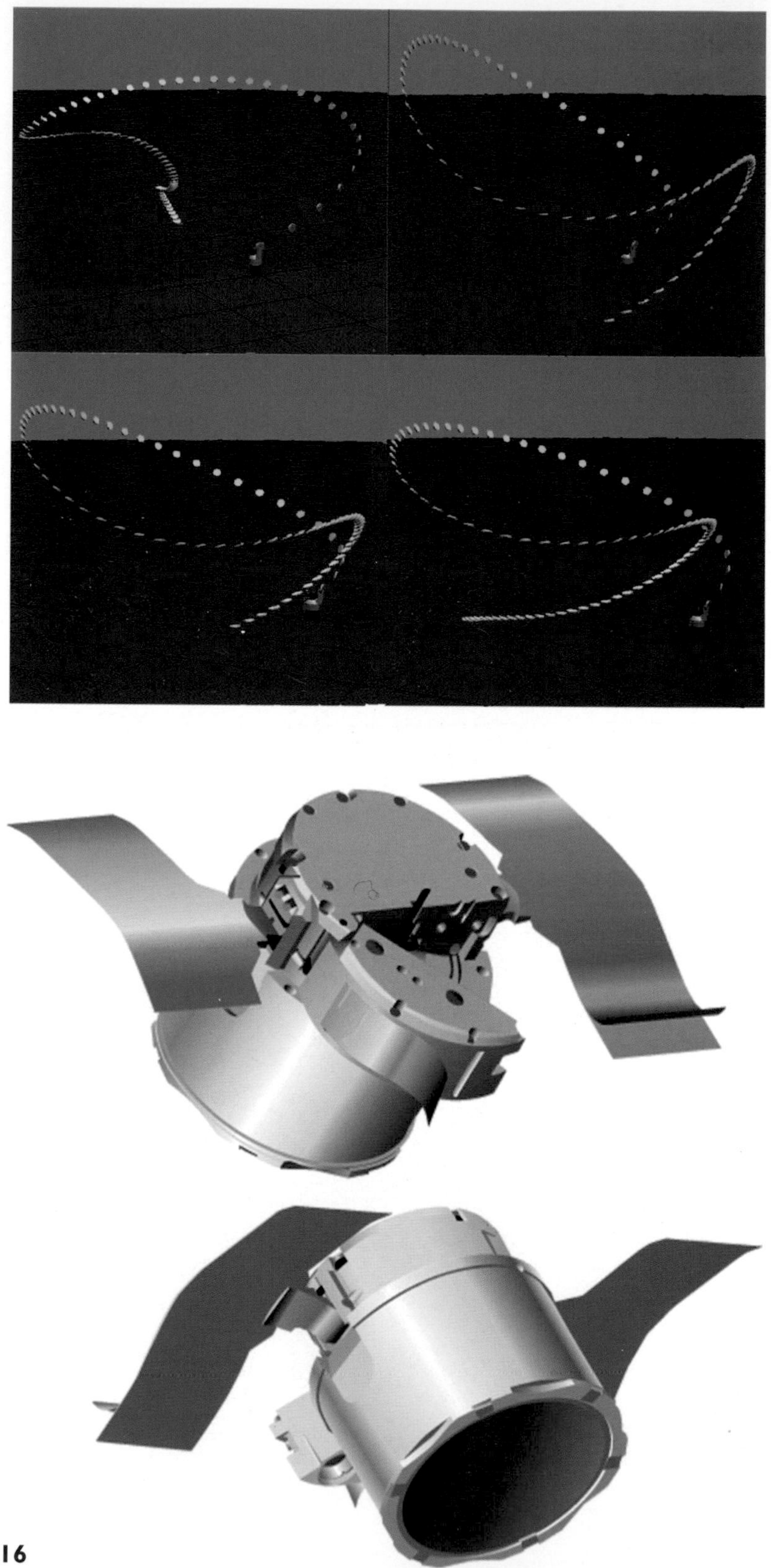

Plate 16

Figure 7.10. Close-up of the *Huygens SM2* model showing the spin vanes. The structure to the left is an attach mechanism for the front shield. Photo by the author.

Figure 7.11. The *Huygens SM2* model with the author at the European Space Operations Centre ESOC in Darmstadt, Germany, just a few days after the real probe arrived on Titan, 1.5 billion kilometers away. The large rectangular box on the top surface is the main parachute container.

After release from the balloon at some 38 km over northern Sweden, the probe accelerated in free-fall before deploying its main parachute and releasing its heat shield. Descending at a fair velocity of over 40 m/s in the thin high-altitude air, the spin vanes caused the probe to spin up, as desired, to a rate of about 20 degrees per second, or about 3 rpm. This equilibrium spin rate slowly declined towards lower altitude, since the descent velocity decreases in the lower, denser air.

The main chute was sized to extract the probe from its wide heat shield, and was actually itself too large to allow *Huygens* to descend in a suitably short time before its communication window ended. Thus the main parachute was detached by a pyrotechnic mechanism which cut the bridle lines, and the probe descended under a smaller "stabilizer" parachute (on a separate swivel of its own). At this point, the probe accelerated to a new terminal velocity, almost twice as fast as under the main chute.

Since the airflow was now streaming past the spin vanes at a higher angle of attack, they worked to spin the probe faster, reaching nearly 40 deg/s before declining at lower altitude.

Because under Earth's conditions (higher gravity and thinner atmosphere than Titan) the impact velocity would be unpleasantly high under the stabilizer chute, the *SM2* test was equipped with a third parachute. This recovery chute was sized to permit a relatively soft landing (10 m/s) and thus possible re-use of the expensive probe model and instrumentation. In fact, the instrumentation recorded the impact and continued to operate, and apart from a bent antenna, the probe was undamaged. This experience gave some hope that the probe might survive landing on a solid surface on Titan, which indeed it did in 2005.

There was no swivel on the recovery parachute. Since there was no such chute on the flight unit sent to Titan, the test did not need to demonstrate a swivel for it. The very different spin behavior under the recovery chute is obvious in the figure: the spin rate winds up to some 30 degrees per second in the direction opposite to that in which the spin vanes were fighting to turn it, then winds down to zero and back up several times, with a period of about 2 minutes. Note that this wind-up/wind-down cycle is not symmetric about zero, since the parachute itself seems to be causing a net negative spin rate.

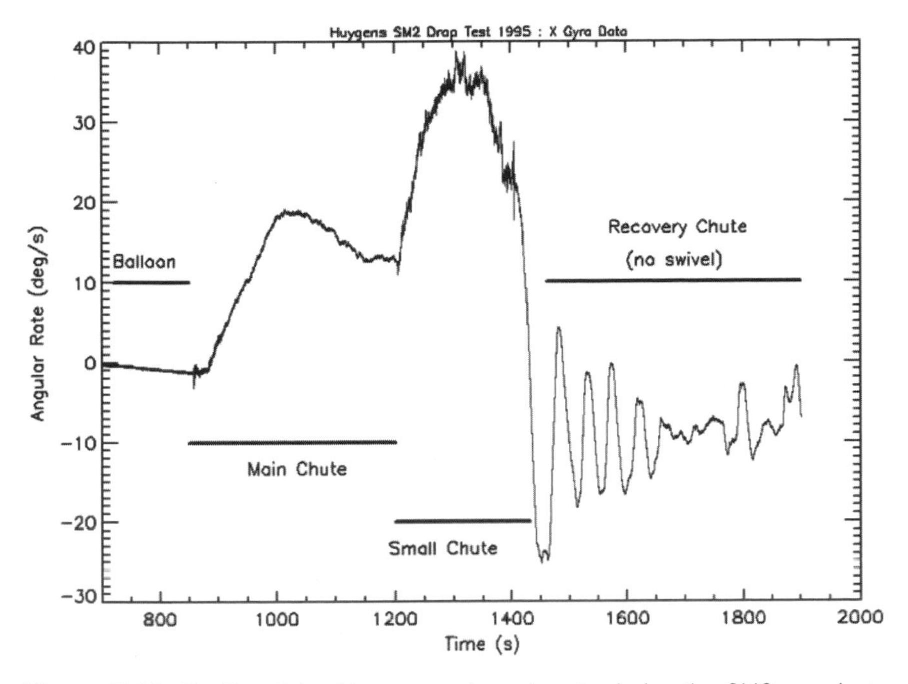

Figure 7.12. Profile of the *Huygens* probe spin rate during the *SM2* parachute drop test on Earth.

On its real mission in space, finally achieved in early 2005, the probe was released towards Titan with a roughly 7 rpm spin, which seems to have been preserved during entry, as expected (unlike *Galileo*).

Huygens was equipped with a multifunction instrument named DISR (Descent Imager and Spectral Radiometer) which as well as measuring light scattering and absorption in Titan's hazy atmosphere, took pictures of the surface. The camera looked down and outwards, from 6 degrees above horizontal to almost vertically down. By taking images at a range of azimuths as the camera was panned around by the probe spin, panoramic mosaics could be constructed.

The instrument used information on the spin rate from the probe's computers, which used a radial accelerometer to estimate the spin rate from centripetal acceleration, and information from a sun sensor. The observation azimuths—most critical for measuring the light scattering around the sun (the solar "aureole")—were not as well placed as had been hoped, due to a combination of circumstances. The sun sensor apparently lost sensitivity at Titan's low temperatures and so was available only in the first part of descent, and there was more tip and tilt motion than had been expected.

Additionally, it seems that for most of the descent, the spin direction was in the direction opposite to that expected. The radial acceleration measurement on board is not able to determine the spin direction, so it was assumed to be in the correct sense. However, apparent spin down to zero and spin up—before a spin-up was expected due to the switch to the smaller parachute—is rather unphysical. What appears to have occurred is that the spin rate declined from its initial 7 rpm down to zero fairly quickly, and the angular acceleration continued in the same sense to reach nearly 10 rpm in the opposite direction to the initial one, then declined in magnitude as the probe descended more slowly in the deeper atmosphere. The cause of this reversed spin torque has not yet been determined at the time of writing.

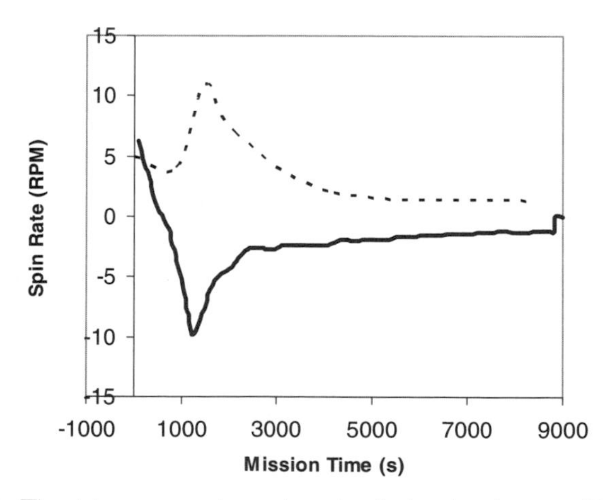

Figure 7.13. The *Huygens* probe spin rate during its descent. The dashed line shows the predicted spin rate profile. The solid line was reconstructed from camera and radio link data after some weeks of analysis on the ground. The source of the torque causing the change in spin direction is not yet understood.

Figure 7.14. An early mosaic of images from the Descent Imager Spectral Radiometer on the *Huygens* probe. The rectangular field of view of the camera was mapped onto the surface of the planet to cover a wide area, although problems with spin sensing led to irregular sampling of azimuths. Note the river channel and coastline to the left. Photo: University of Arizona/ ESA/NASA.

In the event, there were enough mesurements at various azimuths to make a good mosaic of surface images and measure the light in different directions. But the experience underscores the need not only to perform tests before flight, but also to understand and act upon the test results. Also, better instrumentation (e.g. gyroscopes) would have helped reconstruct Huygens' motion.

Spinning Parachute

Although in general spin is a bad thing for parachutes—indeed the phrase "spin parachute" is usually applied to a parachute whose function is to stabilize an aircraft that has entered a spin—there are a few instances where spin is advantageous. One is where spin makes the parachute perform better; another is where it is desired that the payload spins. In this latter category fall certain munitions, as well as probes descending through the atmospheres of other planets.

There are two possible advantages for a parachute to spin about its axis of symmetry. One is gyroscopic stability, a common theme in this book. The spin will help to retard the effects of sideways torques which might otherwise introduce pendulum-type motions.

The more significant effect, however, is on drag performance. The drag coefficient of a parachute is defined relative to its constructed area—the area of the fabric laid out flat. The constructed area usually relates directly to the mass of material required, and is also much easier to measure than the projected area in flight, which can vary. Drag coefficients of 0.5 to 0.6 are usual.

The load on the lines to a parachute tend to cause it to contract, such that the parachute envelope may be somewhat convex. Even in the case of a perfectly hemispherical parachute (not, in fact the best performer, or the most stable), the edge of the chute is at best vertical and so that part of the fabric is not contributing to drag area: the projected diameter of the parachute is $\pi/2$ times less than the constructed area. One way around this problem is to use some sort of rigid or semirigid member to hold the mouth of the parachute open.

The other, rather clever, way is to use centrifugal force to hold the mouth open and give the parachute a flatter profile than it would otherwise have. The gores of the parachute (or "Rotating Flexible Decelerator") are constructed such that there is a radial air gap between them that causes the air loads to make one side bulge out, in some ways resembling an egg beater. As the air spills out, each gore (or rotor blade, in effect) is held at an angle of attack that causes it to move forward around the central axis of the parachute. Usually a central disk of cloth is needed to hold the rotor in the correct shape during inflation.

Figure 7.15. A rotating parachute canopy, constructed from slightly asymmetric gores allowing an "eggbeater" shape. Note the gores do not reach all the way to the center, which is instead blocked by a disk.

In this design (Pepper, 1984), a test rotor with a 24-inch constructed diameter weighed fractionally less than a conventional parachute with a 19-inch diameter (since the rotachute gores have air gaps and so only fill about 2/3 of the possible area, the mass of fabric for a given constructed diameter is less). However, the drag coefficient was twice as high—subsonic values for the rotachute are some 1.0 to 1.2, declining to around 0.6 only at Mach 2. At Mach 2, the chute exhibited the alarming speed of 130 revolutions per second; these rates were such that the swivels had to be replaced between tests. However, the high spin had the desired effect of holding the parachute flat and open. The rotachute showed oscillations of less than 3 degrees.

Some older rotating parachute designs (e.g., see Maydew et al., 1999 for a review of parachute technology) include the Rotafoil, developed by E. Ewing of the Radioplane Company in the 1950s; this is like a conventional circular parachute, but has cut-out slots in the canopy which cause rotation. Another is the Vortex Ring, made of four asymmetrically rigged panels that rotate like autorotating helicopter blades.

SADARM

One application where a parachute is used to spin a payload is in smart weapons. The case in point is the SADARM (Sense and Destroy Armor) munition. This is intended for attacking armored targets such as tanks and artillery. But rather than pound an area with many shells with the expectation that sooner or later the desired targets will be hit, the idea is to deploy a smart submunition that can search an area, find the target within that area, and attack it with a explosively formed projectile warhead. This involves a shaped explosive charge with a metal liner, which when fired blasts down onto the lightly armored upper surface of the target as a jet of molten metal.

Two SADARM submunitions of this type can be launched over 20 km in a 155 mm artillery shell. Six can be packaged in an artillery rocket. After being decelerated over their target, the submunitions deploy a vortex ring parachute (rather like a cluster of asymmetrically rigged parachutes) which allows the system to descend slowly while spinning. This causes an infrared and millimeter-wave radar sensor to scan a spiral pattern on the ground. If a target giving a match to preprogrammed signatures on both sensors is detected, the munition detonates in mid-air and kills its target.

Some initial tests in the early 1990s suffered some problems with mid-air collisions between submunitions, but these were addressed and production began in 1997. The parachute-borne submunition has many conceptual similarities with the samara-wing munitions (STS, "skeet") discussed in chapter 12.

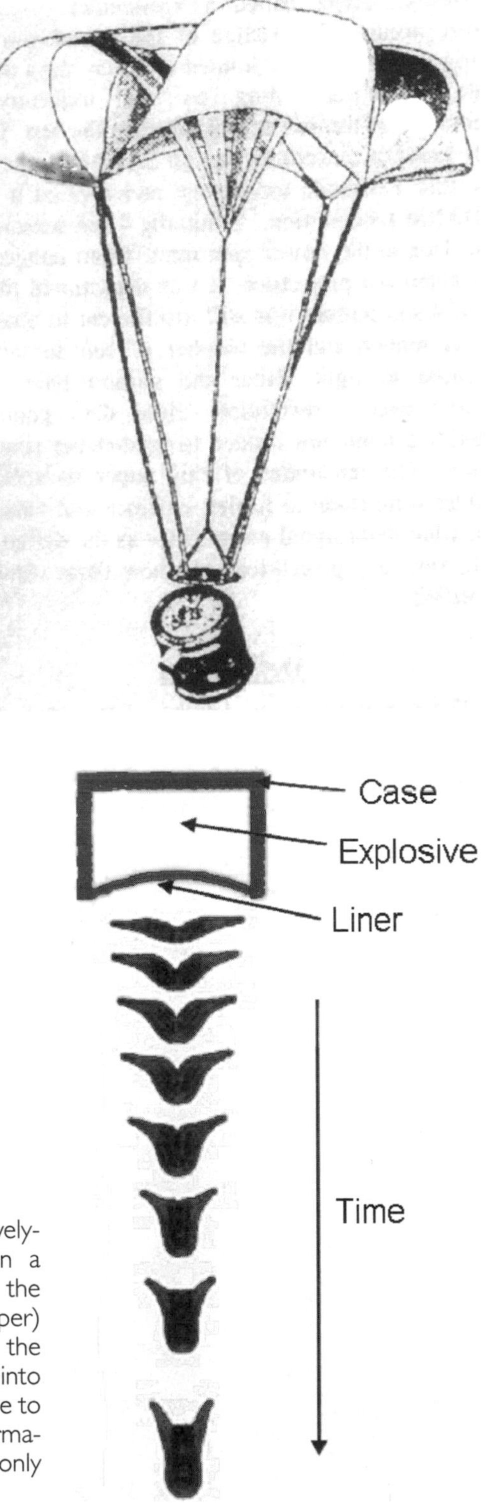

Figure 7.16. Drawing of a parachute-borne smart munition. The rotating parachute cluster causes it to spin about a vertical axis. The suspension system is set such that the payload hangs at an angle and is thus scanned in a conical pattern.

Figure 7.17. Schematic of explosively-formed projectile. The explosive in a cylindrical case detonates, blasting the liner of soft metal (tantalum or copper) downwards. The liner deforms from the dished shape which focuses the blast into a slug traveling at 1000–2000 m/s, able to penetrate 10 cm of armor. The deformation process shown takes place in only 400 microseconds.

Doherr and Synofzik (1986) made a theoretical and wind tunnel study of this application. They studied several types of parachute in a wind tunnel—rotating variants of circular flat, cross, extended skirt, and guide surface chutes (the latter has a rigidizing member to help it hold its shape). The wind tunnel had a horizontal flow; the parachutes were attached to a payload model that was able to spin about a horizontal axis and the spin performance was investigated with a stroboscope. The flat parachute showed the largest sideways oscillations; the guide surface the least.

A key point that Doherr and Synofzik note is that the torque that a parachute can apply to the payload is limited. In order to apply a torque to the payload, the suspension lines must be slightly twisted. But if the torque on the parachute itself is high enough, the chute will rotate relative to the payload to twist the lines excessively, the lines will wrap up, and the parachute will collapse. They present an analytic model of this situation; the result is confirmed by experience, which shows that suspension lines for rotating parachutes need to be short, and the radius at which the lines are attached to the payload needs to be large. If the base diameter of the parachute is ∂_B, and the suspension lines are $0.6\partial_B$ long, the lines can wrap up if they attach at a radius $0.2\partial_B$ or less, but will never wrap (i.e., they can provide an arbitrary amount of torque) if attached at $0.4\partial_B$ or more.

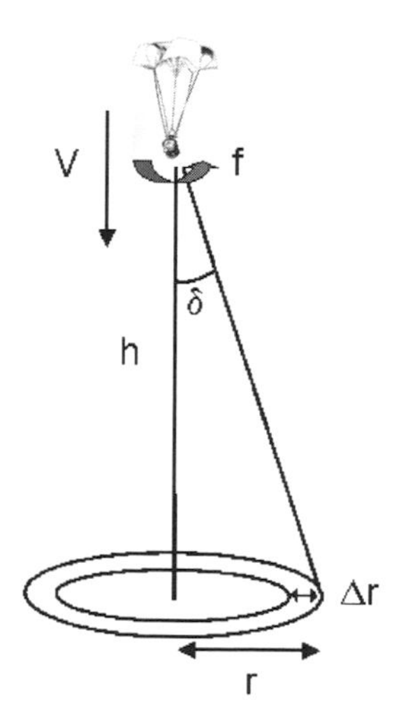

Figure 7.18. Schematic of the scan of a rotating, descending munition. It is desired to keep the interval in horizontal distance between successive scans Δr to a maximum value (say, the characteristic dimension of a tank)—between scans the vehicle will fall a distance $\Delta r\ h/r = \Delta r / \tan(\delta)$, which equals V/f.

They also define a rotor quality number for scanning munitions. This quality R_Q is defined as f/V, where f is the spin rate and V the descent speed. In essence it is the number of revolutions per meter dropped: if the munition falls too quickly, there is not enough time to scan the area. Doherr and Synofzik's data show that the spin rate is proportional to the flight speed and decreases with the parachute diameter, i.e.,

$$f = C_f V / \partial,$$

where C_f is a dimensionless rotor coefficient, which has a value of ~0.36 for extended skirt parachutes; it is the inverse of the advance ratio.

Since the square of the descent rate V is inversely proportional to the square of the parachute diameter ∂ and proportional to the drag coefficient C_d, the quality parameter then collapses to $R_Q = C_f(C_d)^{0.5}$. A typical range is 0.2 to 0.3.

The SADARM munition is released from the shell at high speed (200–400 m/s) and high spin (75–210 Hz). An inflated decelerator (like a ballute—a balloon like parachute) slows the munition down to 60 m/s where the parachute can be released. The parachute spins the payload between 5 and 11 Hz. Measurement of the spin rate, scan angle, and precession of the payload during development in free flight is an interesting problem—sun sensors were used (Pillasch, 1995). Fourier analysis of sun sensor signals showed a 7.06 Hz spin rate and a 0.76 Hz precession rate.

We shall see in a later chapter how the scanning descent objective may be achieved in another way—with a samara wing.

Spinning Parachute-Borne Instrumentation

The region between about 40 km and ~100 km, the Earth's mesosphere and ionosphere, is challenging to study in situ. Indeed, atmospheric scientist Don Hunten has called these altitudes the "ignorasphere," since difficulty of access has led to their being studied rather less than others. These altitudes are too high for balloons, but too low for satellites and thus can only be studied by vehicles on their way up or down.

In one such experiment in 1974, a Super Arcas sounding rocket was used to deploy a scientific payload at 80 km. To maximize the time (just five minutes) spent above 30 km, the 6.5 kg payload was suspended beneath a 5.1 m diameter disk-gap-band (DGB) parachute. The parachute was equipped with four spin vane panels to deflect the airflow through the gap and thus cause it to spin—this was the first ever high-altitude spin parachute. The payload was to measure the atmospheric electric field at these altitudes by the difference in potential on metallized silk electrodes held on the parachute lines. Since offset voltages can appear on electrodes depending on illumination and other factors, the payload had to spin in order to modulate the horizontal component of the field and hence isolate it.

A cross parachute was also considered for this purpose, since it is easy to cause a cross parachute to spin by rigging asymmetrically—making some lines shorter than others. However, the spin rate sensitivity to line length makes the spin rate difficult to predict, and the cross chute could spin so fast that the lines would wrap up (Silbert, 1981). One way of spinning the DGB chute, by attaching canted fins to the band, was abandoned as being impractical to pack in the small volume of the rocket, so deflector vanes were instead installed in the gap. Test drops from a helicopter at 2.5 km showed the low-altitude spin rate was 1.7, 2.7, and 5.4 rpm, depending on the number of vanes installed (2, 4, and 10, respectively.) The effects of varying altitude (since spin rate varies with descent rate, which in turn varies with air density) were studied by three balloon drop tests from 23 km.

The sounding rocket launcher reached a spin rate (see chapter 4) of 18 revolutions per second (1080 rpm). Before the payload was separated, the vehicle was despun to 30 rpm, to assure safe parachute deployment. The spin parachute caused the payload to spin up to 144 rpm—in fact rather higher than planned, since the field sensors were sampled only once per second (the spin was measured with an on-board magnetometer for rates low enough for the telemetry system; additional spin information came from the irregular radiation pattern from the telemetry antenna). During the short spin-up, electric field measurements were successfully obtained in the rocket's frame of reference, although since the dynamics at this point were complex, the field could not be related to an Earth-fixed reference frame. Later in the flight, after the spin rate had declined, the field measurements showed large and irregular variations that are not fully explained.

This particular experience sums up the chapter as a whole: spin control during atmospheric descent is possible, but challenging. It is also surprisingly nontrivial to understand the attitude dynamics of parachute-borne payloads.

References

Bering, E., J. Benbrook, and W. R. Sheldon, Investigation of the electric field below 80 km from a parachute-deployed payload, *Journal of Geophysical Research* 82, 1925–1932, 1977.

Doherr, K.-F., and R. Synofzik, Investigations of rotating parachutes for submunitions, *AIAA 86-2438*, 9th AIAA Aerodynamic Decelerators Conference, Albuquerque, October 7–9, 1986.

Folkner, W. M., Woo, R., and Nandi, S., Ammonia abundance in Jupiter's atmosphere derived from the attenuation of the *Galileo* probe's radio signal, *Journal of Geophysical Research*, Volume 103, Issue E10, pp. 22847–22856.

Jäkel, E., P. Rideau, P. R. Nugteren, and J. Underwood, Drop testing the *Huygens* probe *ESA Bulletin* 85, February 1996.

Kryvoruka, J. K., and T. T. Bramlette, Effect of ablation-induced roll torques on re-entry vehicles, *Journal of Spacecraft and Rockets* 14, 370–375, 1977.

Lanzerotti, L. J., K. Rinnert, D. Carlock, C. Sobeck, and G. Dehmel, Spin rate of *Galileo* probe during descent into the atmosphere of Jupiter, *Journal of Spacecraft and Rockets* 35, 100–102, 1998.

Maydew, R. D., C. W. Peterson, and K. J. Orlik-Rückemann, Design and testing of high performance parachutes, *AGARD-AG-319*, NATO Advisory Group for Aerospace Research and Development, Neuilly sur Seine, France, 1991.

McDevitt, J. B., An exploratory study of the roll behavior of ablating cones, *Journal of Spacecraft and Rockets* 8, 161–169, 1971.

Pepper, W. B., New, high-performance rotating parachute, *AIAA 84-0808*.

Pillasch, D., Data analysis of a spinning payload using sun sensors, *AIAA-95-1543-CP*, 1975.

Seiff, A., Blanchard, R. C., Knight, T. C. D., Schubert, G., Kirk, D. B., Atkinson, D., Mihalov, J. D.. Young, R. E., Wind speeds measured in the deep jovian atmosphere by the *Galileo* probe accelerometers, *Nature* vol. 388, 650–652, 1997.

Silbert, M. N., Deployment of a spin parachute in the altitude region of 260,000 ft, Aerodynamic Decelerator and Balloon Technology Conference, 7th, San Diego, CA, Oct. 21–23, 1981, *AIAA 81-1942*.

Spencer, D. A., R. C. Blanchard, R. D. Braun, P. H. Kallemeyn, and S. W. Thurman, *Mars Pathfinder* entry, descent, and landing reconstruction, *Journal of Spacecraft and Rockets* 36, 357–366, 1998.

Sromovsky, L., A. D. Collard, P. M. Fry, G. S. Orton, M. T. Lemmon, M. G. Tomasko, and R. S Freedman, *Galileo* probe measurements of thermal and solar radiation fluxes in the Jovian atmosphere, *Journal of Geophysical Research* 103, E10, 22,929–22,977, 1998.

Suomi, V. E., L. A Sromovsky, and H. E. Revercomb, Preliminary results of the *Pioneer Venus* small probe net flux radiometer experiment, *Science* 205, 82–85, 1979.

8 Frisbees

Over 300 million Frisbees have been sold worldwide. Once people get the hang of how to throw it, the ability to skim a lightweight object for a hundred meters and have it seemingly hover in midair becomes an addictive pleasure.

People have doubtless flung flat objects around since time immemorial, realizing that spin somehow allowed objects to fly that otherwise could not. The key point is that a flat plate tends to pitch up in flight, and this tendency must be suppressed in order to have sustained flight. The basics are outlined in Schuurman (1990) and Lorenz (2004). This suppression is achieved by some combination of aerodynamic tuning to reduce the pitch-up moment and the application of spin to give gyroscopic stiffness. These are, however, only palliative measures that

Figure 8.1. Stylish and athletic moves characterize the Frisbee sport of Freestyle. Photos courtesy of Larry Imperiale of the Freestyle Players Association: www.freestyle.org.

extend the duration of level flight—simple adjustment of shape and flight parameters cannot keep an object flying forever for the following reason: Spin stabilization only slows the destabilizing precession due to the pitch-up moment—the useful flight time is only a transient interval whose duration is proportional to the spin rate divided by the pitch moment. Of course, if the pitch moment could be made zero, then the spin axis precession would take an infinitely long time. However, it seems impossible to make a flying shape that has a zero pitch moment at all angles of attack, and since the angle of attack will change in flight due to the changing flight speed and flight path angle (due to the actions of gravity, lift, and drag), then sooner or later the pitch moment must be dealt with.

Frisbee History

But long before the problem was thought of in these terms, whatever objects were at hand have been flung with spin. It so happens that one of the more popularized and effective objects were the pie tins of the Connecticut baker William Frisbie. These pie tins were a good size and weight to throw, as students at nearby Yale University found. The deep lip of the tins tended, as we discuss later, to reduce the pitch-up moment, permitting the spin axis to remain stable enough for a flight of a few seconds.

The next major development was the availability of plastics after World War II. Two former Air Force pilots, Warren Francisconi and Walter (Fred) Morrison, saw that plastic would be an ideal material to make a throwing toy. Francisconi and Morrison named their toy a "flying saucer," capitalizing on the recent publicity from UFO sightings in Roswell, New Mexico.

By 1952 sales were not doing well, and Francisconi and Morrison drifted apart. Morrison, however, continued to market flying discs, naming one the Pluto Platter. (The book by Johnson (1975) does not mention Francisconi's role at all; Malafronte's 1998 book gives the two pioneers more or less equal billing.) Morrison patented the disc in his name and teamed up with the Wham-O Manufacturing Co. of California, and the company (with much better marketing abilities than Francisconi and Morrison had) began to sell the discs.

Fortunes improved substantially when a disc player, Ed Headrick, became vice-president at Wham-O, and saw the potential for improving the disc and its sales. He added grooves on the upper surface of the disc (which presumably tripped the boundary layer into turbulence and therefore reduced drag—even such small changes can make performance differences, though probably the key was to differentiate the product enough to patent the new version.) The

"Professional Model Frisbee Disc" received the U.S. Patent 3,359,678 in 1964.

The importance of good marketing cannot be overstated. Although many people were playing with discs, all he did "was offer them a 'Pro' model, white with a black flame painted ring, a gold foil label that said 108 grams, as if anyone cared, and the Olympic rings upside down." The product was promoted by Headrick's forming the International Frisbee Association, setting up the various championship competitions and appearing on TV.

In addition to the millions of Frisbees just tossed between friends to pass the time in a vaguely athletic sort of way, several specific sports have developed, notably Guts, Freestyle, Ultimate, and Disc Golf.

Guts is a game in which two opposing teams take turns throwing the disc at each other, the goal being to have the disc hit the ground in a designated zone without being caught. Freestyle is more of a demonstration sport like gymnastics, with exotic and contorted throws, catches, and juggles evaluated for difficulty, precision, and artistic impression.

The game of Ultimate, a team passing game with similarities to basketball and American football, has become a popular sport, and is featured in the World Games. The rules were developed by high school students in 1968, being refined somewhat in the following few years. In essence, a team must work the Frisbee forward across a 70-yard long field, 40 yards across, by passing from one team member to the other. If possession is lost, either by the disc going out of bounds, falling to the ground without being caught, or being caught by a member of the opposing team, the opposing team takes possession. A point is scored when the disc is caught in the endzone.

It is of course natural to do "target practice" with a disc, and a sequence of targets makes for a golf-like game, with the aim being to hit the targets in as few throws as possible. The practicalities of a target that gives unambiguous indication of a "hit" without damaging the disc led Headrick to patent a Disc Pole Hole, a device which could catch a Frisbee. The device consists of a frame supported by a pole: ten chains hang down from the frame, forming a paraboloid of revolution. This paraboloid sits above a wide basket. The chains absorb the momentum of a correctly thrown disc and allow it to fall into the basket (without the chains, a disc would typically bounce off the pole, making scoring near-impossible). The first Disc Golf course was set up in Pasadena, California (in fact rather close to NASA's Jet Propulsion Laboratory). A large range of different golf discs are available, with their weight and shape optimized for different throw ranges and wind conditions. According to the U.S. Professional Disc Golf Association, there are over 3 million regular players of disc golf, with several hundred tournaments per year.

Participation in Frisbee sports is not even confined to human beings. The TV sports network ESPN has begun to broadcast "Hot Zone," a competition sport where a player throws a Frisbee to be caught in a specified zone by a dog (often a sheepdog breed).

In terms of exploiting the widest range of aerodynamic properties of the disc, Ultimate is arguably the key sport. A thrower must toss to a teammate while avoiding interception, and therefore curved flights are essential. Sometimes the thrower may be blocked by an opposing player and thus must use an exotic throw, such as the overhead "hammer" where the disc is thrown over the shoulder in a vertical orientation, to roll onto its back and fly at near –90 degrees angle of attack. The catcher must anticipate how long the disc may hang in the air, and especially any turns it may make towards the end of its flight.

Innumerable variations on the Frisbee theme have been made—discs with flashing lights, discs with ropes attached so dogs can pick them up easily, inflatable discs, and so on. But in fundamental terms, the simple—albeit cleverly shaped—plastic disc seems here to stay.

Figure 8.2. In the sport of Ultimate Frisbee, players must throw discs to be caught by their teammates without being intercepted by the other team. This nominally noncontact sport requires careful throws that exploit the disc's unusual aerodynamic properties, as well as some athletic catches. Photo courtesy Andrew Davis www.freeheelimages.com.

Mechanics of Frisbee Flight: Wind Tunnel Measurements

As often seems to be the case, it was fairly late in the history of these objects that they began to be studied scientifically. In addition to studying the basic aerodynamic parameters, these investigations have tried to grapple with the possibility that the spin rate may affect not only the gyrodynamics, but also the aerodynamics. These aspects have assumed new importance with the prospect that a controllable drone or Unmanned Aerial Vehicle (UAV) might be patterned on a Frisbee: the behavior of control surfaces such as flaps would need a better understanding of the flow over the disc.

As far as just throwing a disc goes, the basic mechanical point is that a Frisbee is just a low-aspect ratio wing—in that sense its lift and drag can be considered conventionally. Admittedly, it is a wing that is sometimes deliberately operated even at –90 degrees angle of attack, but even then—in common with many low-aspect ratio shapes—its behavior is predictable.

The key is the pitch moment and how to mitigate its effects. The conventional Frisbee does this in two ways. First, the deep lip reduces the pitch moment to manageable values. Secondly, the thickness of the plastic in a Frisbee is adjusted across the disc, such that much of the disc's mass is concentrated at the edge, to make a thick lip. This has the effect of maximizing the moment of inertia, making the Frisbee like a flywheel. The precession rate is equal to the pitch moment divided by the moment of inertia and spin angular velocity. Keeping the precession down to a few degrees over the flight duration of a couple of seconds is all that is needed.

Figure 8.3. Schematic of the behavior of an object with pitch-up moment.

The earliest documented wind tunnel measurements of Frisbee-type vehicles appear to be those of Stilley and Carstens (1972), who reported force and pitch moment coefficients for a nonspinning disc, and asserted that the forces at least were unaffected by the spin rate. (Their interest was in the possible use of a spinning disc configuration to loft a flare.)

Nakamura and Fukamachi (1991) performed smoke flow-visualization experiments on a Frisbee at low flow velocity in a wind tunnel (~1 m/s). The disc (a conventional, although small, Frisbee) was spun with a motor at up to 3 times per second, yielding an advance ratio of up to 2.26. The smoke indicated the presence of a pair of downstream longitudinal vortices (just like those behind a conventional aircraft) which create a downwash and thus a lift force. These investigators also perceived an asymmetry in the vortex pattern due to the disc's spin, and also suggested that the disc spin increased the intensity of the downwash (implying that the lift force may be augmented by spin). While this work may be the first investigation of these phenomena, it is important to bear in mind the low flow velocity (~20 times smaller than typical flights); the effect of rotation in these experiments may have been disproportionate.

Yasuda (1999) measured lift and drag coefficients of a flying disc for various flow speeds and spin rates. He additionally performed a few free-flight measurements on the disc (with the disc flying a short distance indoors in the field of view of a camera) and wind tunnel measurements on a flat disc. His free-flight measurements on a conventional disc show that typical flight speeds are 8–13 m/s and rotation rates of 300–600 rpm (5–10 revolutions per second) and the angle of attack was typically 5–20 degrees. The most common values for these parameters were about 10.5 m/s, 400 rpm, and 10 degrees, respectively.

The flat disc had a zero lift coefficient at zero angle of attack, and a lift curve slope between 0 and 25 degrees of 0.8/25. The Frisbee had a slight lift ($C_L \sim 0.1$) at zero angle of attack, and a lift curve slope of ~1/25.

The Frisbee paid a price for its higher lift: its drag was commensurately higher. The flat plate had a drag coefficient at zero angle of attack of 0.03 and at 25 degrees of 0.4; the corresponding figures for the Frisbee were 0.1 and 0.55. (The drag curves are parabolic, as might be expected for a fixed skin friction drag to which an induced drag proportional to the square of the lift coefficient is added.) Yasuda notes that the lift : drag ratio of a flat plate is superior to that of the Frisbee. No significant dependence of these coefficients on rotation rate between 300 and 600 rpm was noted.

Higuchi et al. (2000) performed smoke wire flow visualization and PIV (particle image velocimetry) measurements, together with oil flow measurements of flow attachment on the disc surface. They used a cambered golf disc, with and without rotation and (for the most part) a representative flight speed of 8 m/s, and studied the downstream vortex structure and flow attachment in some detail.

To date, the most comprehensive series of experiments have been conducted by Jonathan Potts and William Crowther at Manchester University in the UK, as part of the Ph.D. research of the former. One aim of the research was to explore the possibilities of control surfaces on a disc wing.

In addition to measuring lift, drag, and pitch moment at zero spin for the classic Frisbee shape, a flat plate, and an intermediate shape, these workers measured these coefficients as well as side-force and roll moment coefficients for the Frisbee shape at a range of angles of attack and spin rates. These coefficients will be discussed in a later section.

Additionally, Potts and Crowther performed pressure distribution measurements on a nonspinning disc, smoke wire flow visualization, and oil flow surface stress visualizations. (They performed these on the regular Frisbee shape, and one with candidate control surfaces.)

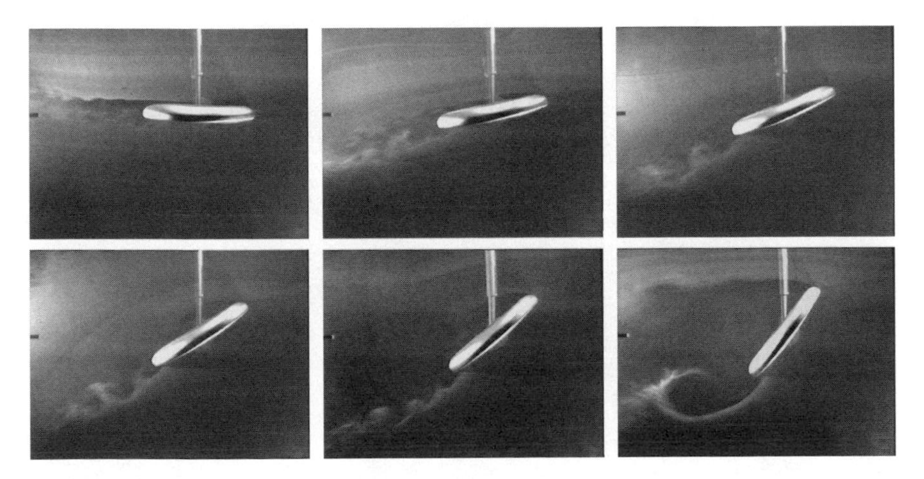

Figure 8.4. Smoke flow visualization of a nonspinning Frisbee model at increasing angle of attack (0–50° in 10° increments). Separated flow on the upper surface (stall) and strong vortex shedding into the wake is evident at high angles of attack.

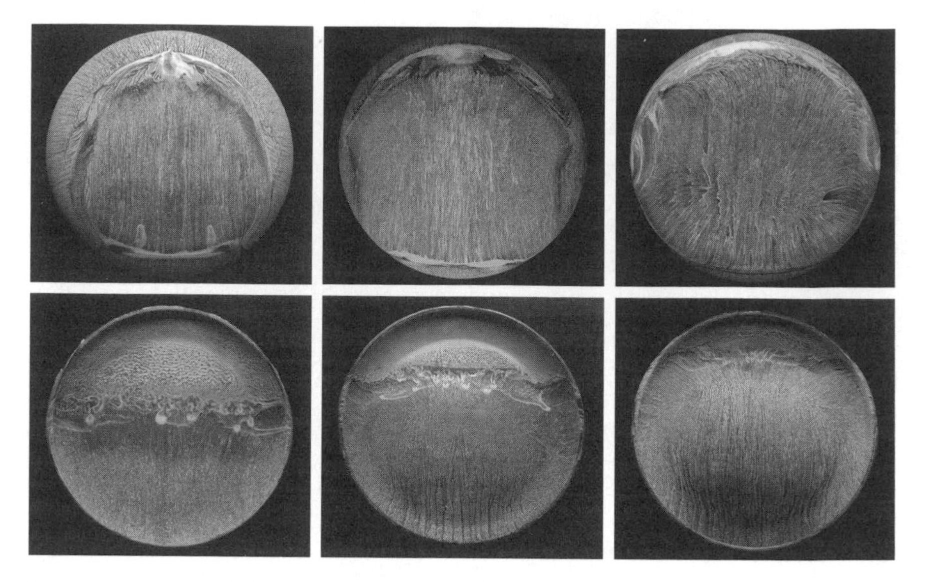

Figure 8.5. Upper (top) and lower (bottom) surface paint flow photographs illustrating the surface stress on a Frisbee at angles of attack of 5°, 15°, and 25°. Flow speed is 15 m/s. Photo courtesy of Richard Crowther and Jonny Potts.

In addition to these papers cited above, wind tunnel measurements on Frisbees appear to be a perennial research topic for undergraduate education. The quality of these measurements, and the rigor of their reporting, can be variable, and these reports are difficult to obtain. The determined investigator should consult the reference list of Potts and Crowther (2002) for at least a partial list of such reports.

Figure 8.6. Perspective view of fluorescent paint flow. Notice how the surface flow diverges behind the leading edge, where the separation bubble reattaches. Figure by Potts and Crowther, used with permission.

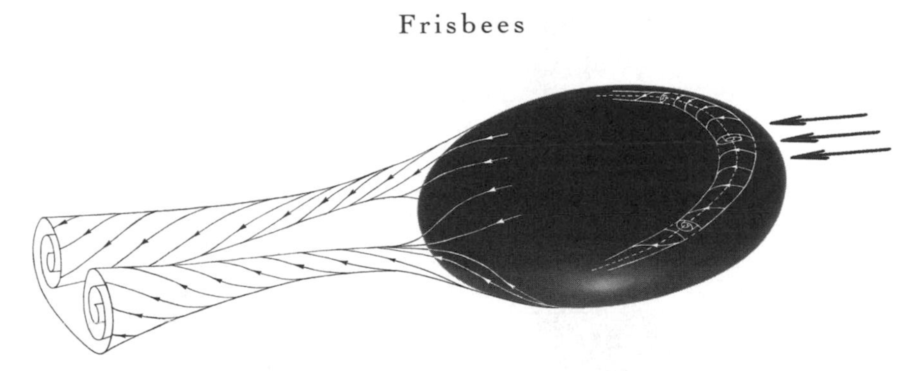

Figure 8.7. Sketch of the flow circulation on a Frisbee at a typical flight condition. Two trailing vortices like those on conventional aircraft are shown, together with a small circulation bubble associated with the suction peak just behind the leading edge. Figure by Potts and Crowther, used with permission.

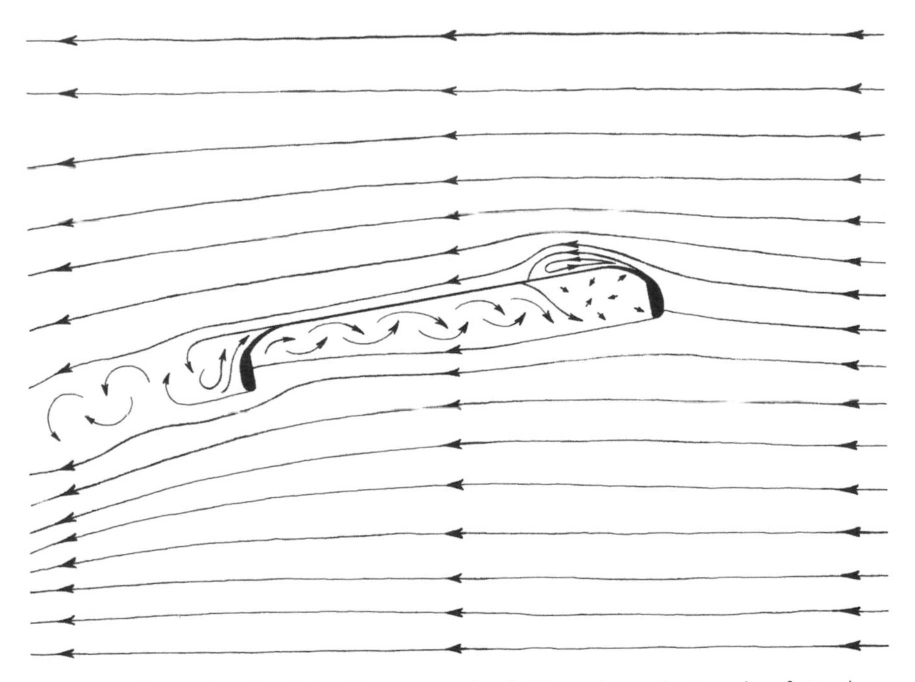

Figure 8.8. A sketch of the flow around a Frisbee at modest angle of attack, as deduced from flow visualizations. Notice how the trailing side of the lip "catches" the airflow to cause a pitch-down. Figure by Potts and Crowther, used with permission.

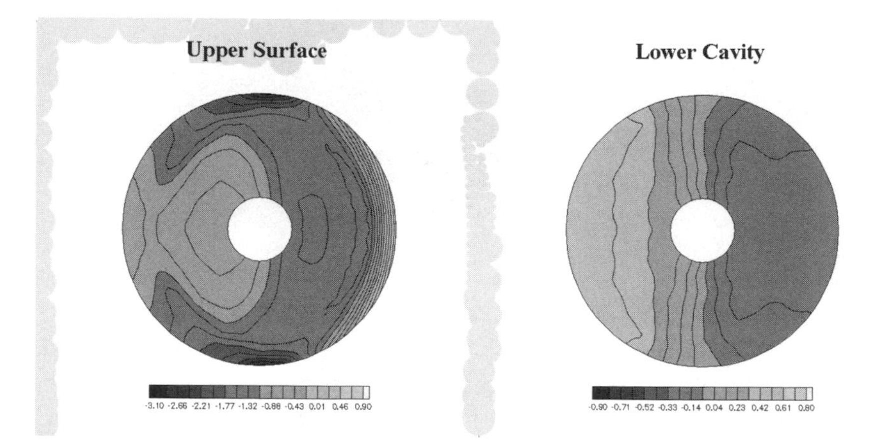

Figure 8.9. Pressure distribution on a Frisbee's upper and lower surface at 14 m/s and 15° angle of attack. Figure by Potts and Crowther, used with permission.

Computational Fluid Dynamic Studies

In the last decade or so it has become practicable to substitute calculations (numerical solutions of the Navier-Stokes equations) for wind tunnel measurements.

One undergraduate report (at the time of writing, an electronic file was available at Sarah Hummel's website) is that of Dankowsky and Cohanim at Iowa State University in 2002. These students performed some wind tunnel measurements (NB of a "golf disc" with a flat rim, rather than the classical Frisbee configuration) as well as simple "panel method" aerodynamic calculations.

A more elaborate CFD (computational fluid dynamics) study was undertaken by Axel Rohde in connection with his Ph.D. thesis. This study explored the flow around a disc (again a disc golf type, although a more conventional cross-section than that just above.) It should be noted that the thrust of the thesis was the development of a CFD solution method (a code for implementing this method is available on the web at www.microcfd.com, at which site the Ph.D. thesis can also be downloaded) rather than the study of Frisbee aerodynamics per se, and the flight parameters explored in this work are not those one might view in the park or on a beach—specifically the disc is a spinning oblate spheroid and more particularly the flight Mach number is 0.5!

CFD remains an area with considerable unrealized potential in Frisbee studies. However, as with wind tunnel experiments, spin introduces significant complications.

Free Flight Studies

Another method of determining the aerodynamic properties of an object is by observing its trajectory in free flight, and determining by simulation what coefficients are consistent with the observed flight path. (This is the principal technique applied in studying boomerang flight—see chapter 11.) To obtain a full set of coefficients with reasonable confidence, a number of different flights must be made, and it is possible—if not likely—that some regions in parameter space cannot easily be explored this way. Stilley and Carstens also matched the trajectories of flying discs, albeit rather stubbier ones than conventional Frisbees, by launching them off a cliff and observing with cine film!

Sarah Hummel and Mont Hubbard at UC Davis in California employed the video approach, using high-speed video cameras to track LED markers on a thrown Frisbee. (Yasuda et al. used video to determine the typical free-flight parameters of a disc, but used wind tunnel measurements to evaluate the coefficients at those parameters.) The sequence of position measurements was then used to evaluate the aerodynamic coefficients by progressively adjusting an assumed set of coefficients and forward-simulating the flight until the squared differences between the simulated and observed positions were minimized. The same technique has been applied in ballistic range tests of hypersonic re-entry vehicles at NASA Ames. Scale models of capsules and probes are shot down a partly evacuated tunnel along which a series of cameras are mounted: measuring the position and orientation of the vehicle at each spot allows the aerodynamic coefficients to be measured.

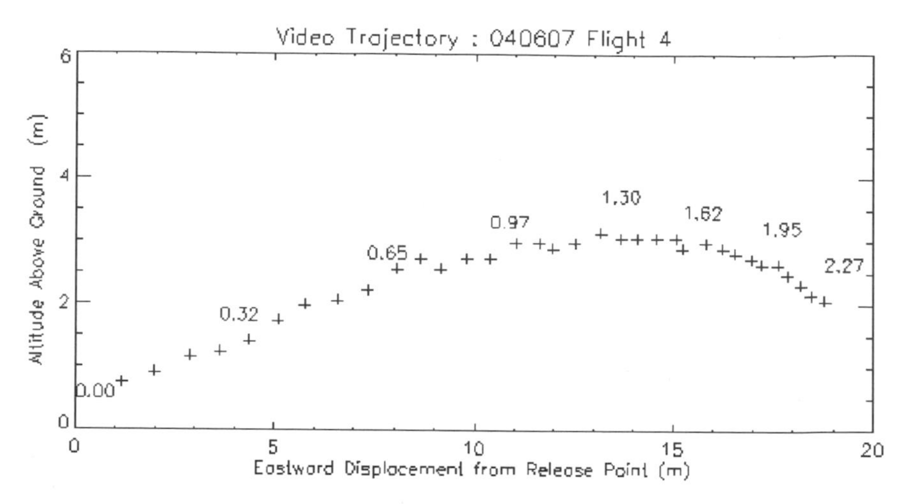

Figure 8.10. A Frisbee flight path obtained by digitizing the location of the disc in a video sequence obtained with a conventional camcorder from ~20 m away. The numbers along the trajectory are the time in seconds since launch.

Aerodynamic Coefficients of Frisbees

A useful and instructive comparison can be made between a flat plate and a Frisbee. Indeed, Potts and Crowther make measurements of both. Let us first consider drag. The drag coefficient is the drag force normalized with respect to dynamic pressure ($0.5\rho V^2$) and the planform area of the disc. Since at low incidence angles the area of the disc projected into the direction of flow is very small (they used a plate with a thickness : chord ratio of 0.01), it follows that a flat plate will have a very low drag coefficient, ~0.02. On the other hand, the Frisbee, with its deep lip (thickness : chord ratio of 0.14) has a much larger area projected into the flow, and its drag coefficient at zero angle of attack is therefore considerably larger (~0.1). The Frisbee maintains a more or less constant offset of 0.1 above the value for a flat plate. This in turn has a parabolic form with respect to angle of attack, owing to the combination of a more or less constant skin friction drag term and the induced drag term, which is proportional to the square of lift coefficient.

While a flat plate has zero lift at zero angle of attack, and a lift coefficient that increases with a slope of ~0.05/degree, the Frisbee, having a cambered shape, develops appreciable lift at zero angle of attack ($C_{L0} \sim 0.3$)—its lift curve slope is similar.

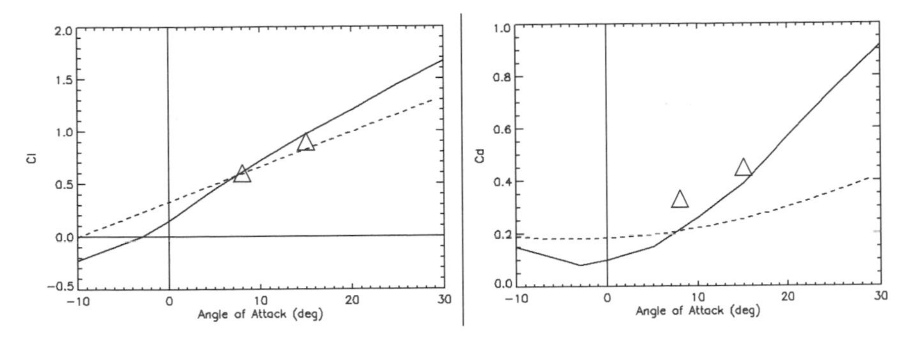

Figure 8.11. Lift and drag coefficients of a Frisbee. Solid line is wind tunnel results from Potts and Crowther; dashed line is from video measurements by Hummel; triangles are free-flight acceleration data from Lorenz (see later).

The major difference between the Frisbee and flat plate is in the pitch moment coefficient. While this is zero for a flat plate at zero angle of attack (which is not a useful flying condition, since a flat plat develops no lift at this angle!), it rises steeply to ~0.12 at 10 degrees. Because the Frisbee's trailing lip "catches" the underside airflow and tries to flip the disk forward, the pitch-up tendency of the lift-producing suction on the leading half of the upper surface is largely compensated. Its pitch moment coefficient is slightly negative at low incidence and is zero (i.e., the disc flies in a trimmed condition) at an angle of attack of about 8 degrees. Over the large range of angle of attack of −10 to +15 degrees, the coefficient varies only between −0.02 and +0.02.

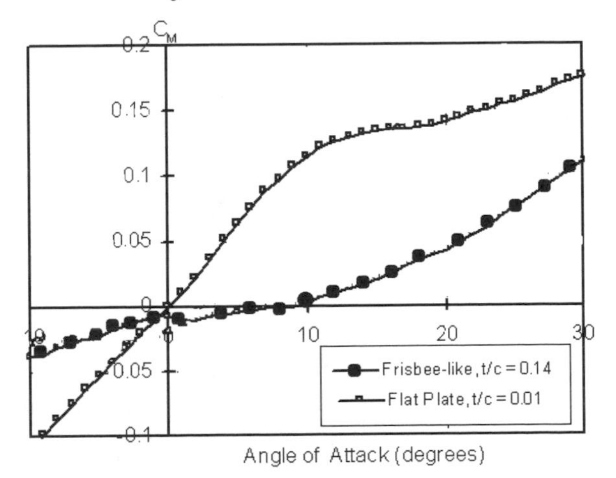

Figure 8.12. Pitch coefficient measurements by Potts and Crowther showing that a Frisbee-like shape has a much lower pitch moment than a flat plate. Indeed, the Frisbee pitch moment coefficient is zero at around 8 degrees.

The astute reader will realize that, while perhaps not useful in a game of Ultimate, the existence of a trimmed position (pitch moment coefficient CM = 0) permits the possibility of a stable glide. If the disc is flying downwards at a speed (dictated by the lift coefficient at the trimmed condition) such that drag is balanced by the forward component of weight, then the speed will remain constant. However, although the zero pitch moment means the disc will not roll, the roll moment is not zero, and so the spin axis will be slowly precessed forward or back, changing the angle of attack.

Hummel has pointed out the role of the sign change in pitch moment in causing the sometimes serpentine (S-shaped) flight of Frisbees. When thrown fast at low angle of attack, the pitch moment is slightly negative and causes the Frisbee to very slowly veer to the right. However, as the disc's speed falls off, its lift no longer balances weight and it falls faster downwards, increasing the angle of attack. When the angle of attack has increased beyond 9 degrees, the pitch moment becomes positive and increases rapidly. This leads to the often-observed left curve at the end of a flight.

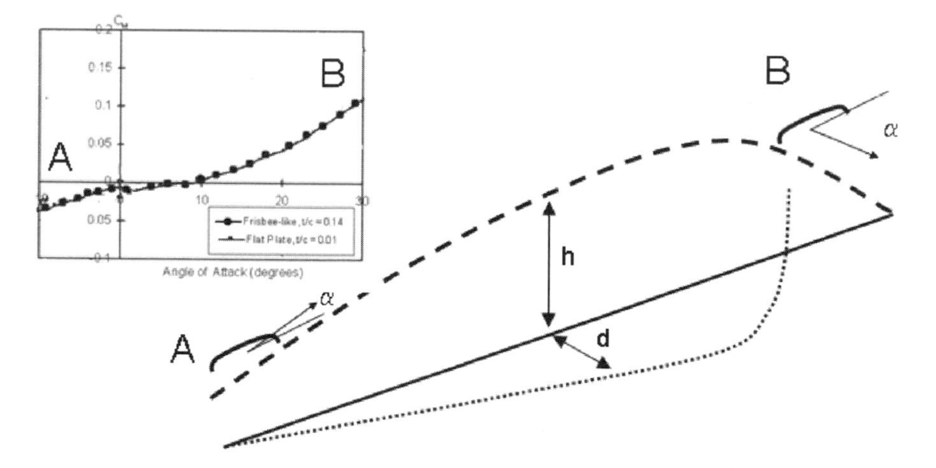

Figure 8.13. Why a Frisbee "hooks" to the left at the end of its flight. At the beginning (A) the flight path (dashed line) is shallowly upwards and the angle of attack α is small or negative, with a small and negative pitch moment causing a slight curve to the right as indicated by dotted line groundtrack. Later, however, at (B) the Frisbee is falling and has a large positive angle of attack and so a large positive pitch moment, causing the veer to the left.

Potts and Crowther also study the side-force coefficient (which might be thought of as due to the Robins—Magnus force, although in reality it is rather more complicated, since most of the boundary layer develops over the flat surface of the disk, rather than its somewhat cylindrical lip) and the roll moment.

The side-force coefficient is not strongly variable over the range of angles of attack studied (–5 to 15 degrees). It does vary, as one might expect, with spin rate. For low values of advance ratio AR (< 0.5, at an airspeed of 20 m/s) the coefficient is just slightly positive (0.02). However, for more rapid spin, the coefficient increases—at AR = 0.69, $C_s = 0.04—0.05$, and for AR = 1.04, $C_s \sim 0.8$. To first order, then, these data show that the side-force coefficient is proportional to advance ratio; a reasonable expectation is that the coefficient is in fact directly proportional to the tip speed, although this parameter was not varied independently in these tests.

Although the lift and drag coefficients were not appreciably affected by spin, the pitch moment did become more negative (by 0.01 —almost a doubling) at 0–10 degrees angle of attack as the spin rate was increased from AR = 0 to 1.

The roll moment coefficient was also determined. This was almost zero (within 0.002 of zero) for low spin rates and more or less constant with angle of attack over the range –5 to 15 degrees. However, the higher aspect ratio data showed a significant roll moment—$C_M \sim -0.006$ for advance ratio AR = 0.7 and $C_M \sim -0.012$ for AR = 1: in both cases the moment coefficient increased in value with a slope of about 0.0006/degree. Although measuring and understanding these coefficients is important in considering long-duration flight stability of powered or guided disc-UAVs, whether these more subtle effects can be exploited in a controlled and conscious fashion in a Frisbee throw is not yet clear.

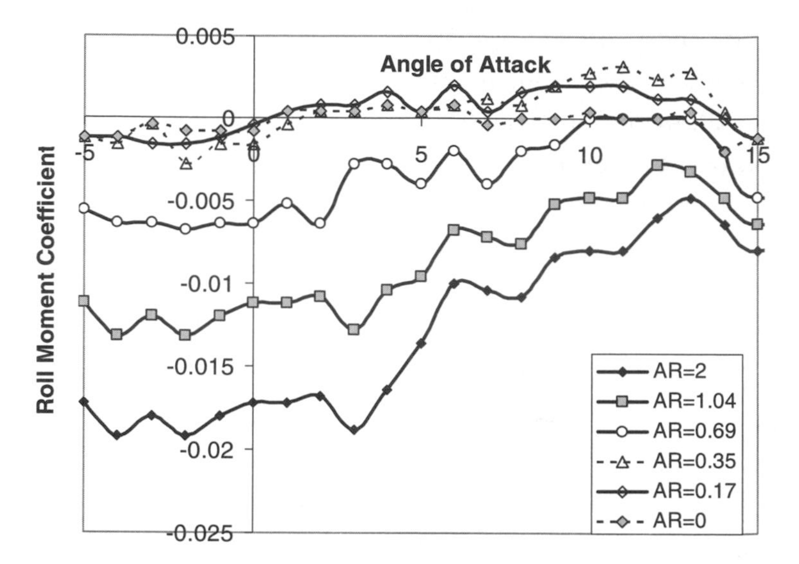

Figure 8.14. The effect of spin on a Frisbee's roll moment coefficient from the experiments of Potts and Crowther. Below advance ratios (AR) of about 0.5, the coefficient is small, while for AR of 1 and higher the moment becomes quite significant.

Instrumented Free-Flight Experiments

My own investigations into Frisbee dynamics (Lorenz, 2005) have centered on using instrumented discs to record accelerations and attitude motions during flight. Data is acquired using microcontrollers mounted on or in the disc from sensors which include sun sensors and magnetometers to measure attitude (calibrated by mounting the Frisbee on a "lazy susan" turntable, in turn set up on the precision angle mount of an 8-inch telescope), accelerometers, and other sensors like microphones and pressure sensors. The appendix to this book gives further details.

Note that to recover aerodynamic coefficients from in-flight accelerations, it is necessary to also measure the flight speed and flight path angle so that the angle of attack can be reconstructed from the attitude. Even a crude video record such as that in Figure 8.10 is adequate for meaningful results to be calculated (triangles in Figure 8.11.)

Among the interesting phenomena identified in these studies is the prominent existence of nutation in the early part of the throw. A good

Figure 8.15. Frisbee underside with electronic components mounted. Most are glued onto the base or rim of the disc, and covered with clear adhesive tape to minimize abrasion damage and airflow disruption. The heaviest components, the batteries, are mounted in cavities milled into the rim to maximize the moment of inertia.

throw will avoid exciting nutation, which seems to substantially increase drag. Hummel's video work has also identified this, although whether it is damped by aerodynamic effects, or structural dissipation, remains to be determined. It can be seen in some photographs of hard Frisbee throws that the disc becomes visibly deformed by inertial loads—the disc is held only at one edge, and to reach flight speeds of ~20 m/s in a stroke of only a meter or so requires 20 g or so of acceleration. Consider half the disc (90 grams) being accelerated at this rate as needing a force of 20 N: since this is roughly equivalent to hanging a 2 kg weight at the edge of the disc, one can readily imagine a transient deformation that might excite nutation.

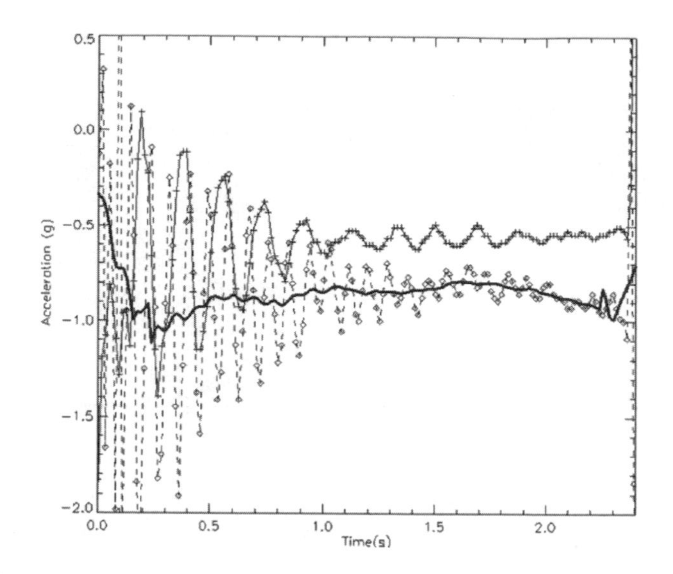

Figure 8.16. Accelerations measured during a conventional Frisbee flight. Solid line with crosses is the radial acceleration, with a centripetal component upon which a once per revolution drag is superimposed. The dashed-diamond curve is the axial acceleration. This shows a twice per revolution variation, characteristic of nutation. The thick line is a smoothed version of the axial data, showing how lift balances about 90% of the weight throughout the flight.

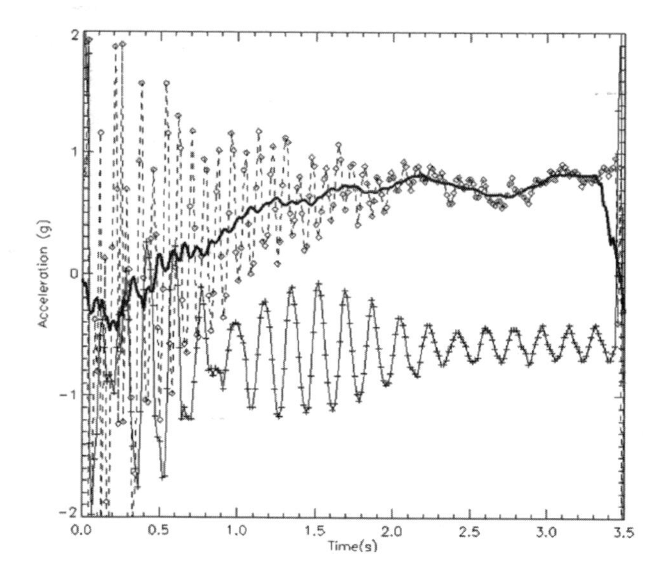

Figure 8.17. Same as Figure 8.16 but for a "hammer" throw. In this throw the Frisbee is thrown in a roughly vertical orientation at a slightly negative angle of attack. The pitch moment causes the disc to roll onto its back, giving the ~+1 g axial acceleration. As the disc turns over, the lift and drag components in the radial direction cancel out temporarily. The spin modulation on the radial component vanishes temporarily before growing to a maximum and then falling again.

In-flight measurements offer the prospect of measuring flow properties such as pressure on the rotating disc. Pressure distributions can be measured in the wind tunnel (Figure 8.9 shows data from Potts and Crowther), but because these measurements use little pipes to draw the pressure from the disc to an array of pressure sensors, it is impossible to spin the model. Free-fright experiments could explore how spin affects the flow separation near stall—trial measurements with just a small microphone show how as the angle of attack increases, the pressure fluctuations on the disc become larger even as the flight speed decreases towards the end of the flight.

A control surface, such as a flap, would have little useful effect on a Frisbee's flight were it to be simply fixed onto the disc. As the disc spins around, the control effect would vary or even reverse, and the spin-averaged effect would be small. However, if on-board sensors were used to trigger a fast-acting flap at a particular phase of rotation, the prospect of a maneuverable Frisbee can be envisaged. This might simply involve some stability augmentation—say to suppress the hook at the end of the flight. But much more appealing ideas become possible—a Frisbee with a heat sensor to detect a player, such that the disc tries to avoid being caught!

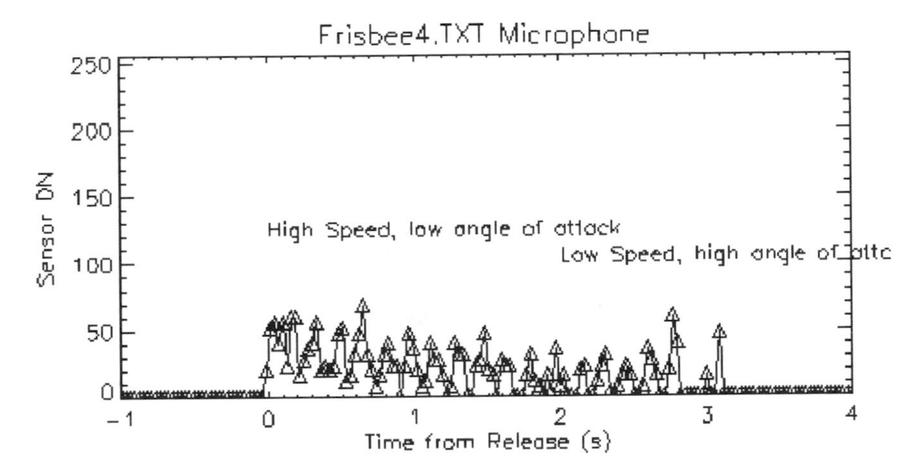

Figure 8.18. Signal from a small microphone on the upper surface of a Frisbee. The signal, which presumably corresponds to turbulent fluctuations in pressure on the upper surface, is spin-modulated at 6 Hz or so at the beginning of the flight. The mean signal falls as the disc slows down, but oscillations increase towards the end of the flight as the angle of attack increases.

Biomechanics of the Frisbee Throw

The challenge in the Frisbee throw is that the overall flight is very sensitive to the initial parameters—small variations in angle of attack can lead to very different flights. (Were this not the case, Frisbee might lose much of its appeal.) Consequently, it can be frustrating to learn.

As performance in sports becomes ever more important, scientific methods can be applied to understand how the throw is executed and how it could be improved. This is not to say that describing to a person what the velocity and angle history he or she should apply to the disc will actually allow them to execute the throw (neuromuscular control in humans is not, like the kinematics of a robot, specified as a set of deterministic commands), but it does give some insight into the technique.

This sort of biomechanical study formed another part of Sarah Hummel's work. It involved the construction of a mathematical musculoskeletal model of the Frisbee throw. The thrower is modeled as a kinetic chain of rigid elements—twist of the torso, and angular motion around the shoulder, elbow, and wrist cause the hand/disc to swing through space at high speed before the disc is released.

High-speed (180 frames per second) video was obtained with four cameras to track the motion of reflective markers mounted on a test subject who threw the disc. Torso twist defines three phases of the throw—wind-up, acceleration, and follow-through. Wind-up refers to the left twist before the throw.

In the acceleration phase, the arm becomes uncoiled, and the torso twists right and bends forward as the player shifts weight from the left to the right foot. A typical history of the various angles is shown in Figure 8.20.

By introducing the masses and moments of inertia of the various kinetic segments, the torques and energies exerted by each joint can be determined—the horizontal adduction of the humerus is the prime source of energy at release, with a power of some 100 W. The wrist flick only contributes around 8 W.

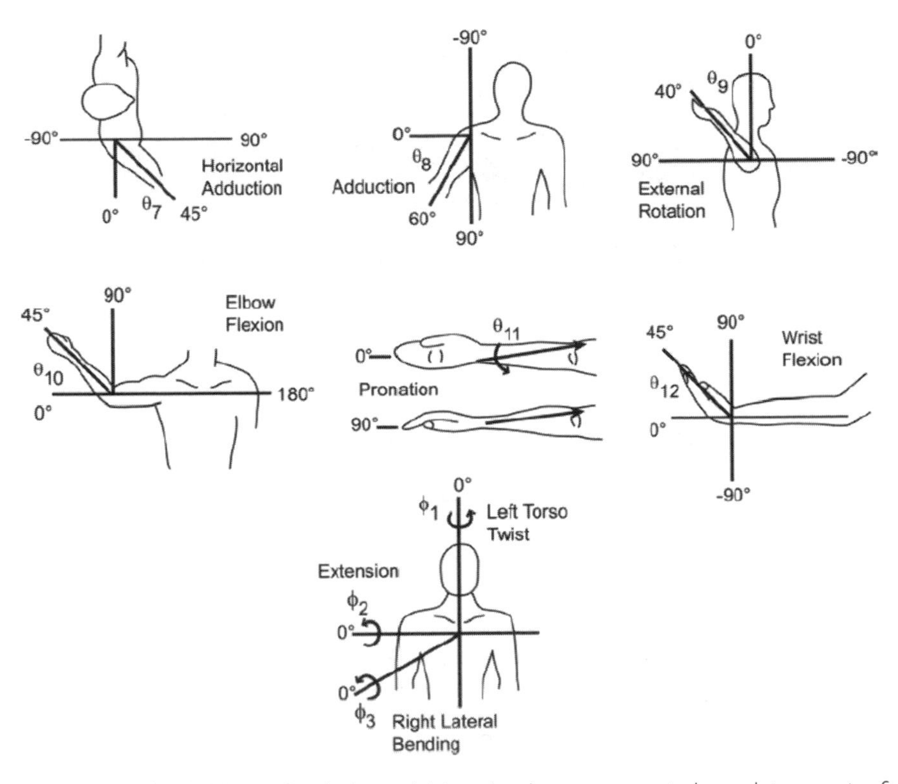

Figure 8.19. A biomechanical model breaks the movement down into a set of joint rotations. Since the bones between the joints are of fixed length, the set of joint angles defines the state of the system. The known masses of the various arm etc. segments allow the work performed in each angular acceleration to be calculated. Figure courtesy of Sarah Hummel.

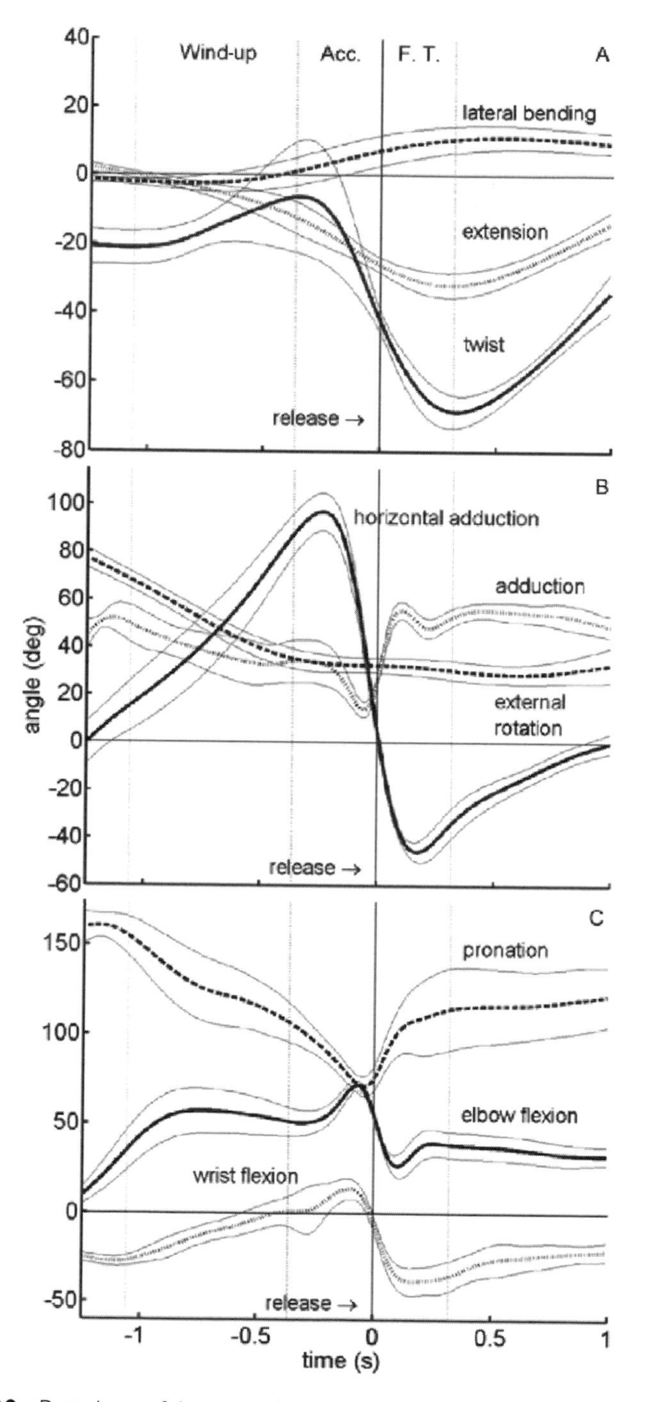

Figure 8.20. Rotations of the torso, humerus and ulna measured from high-speed video by Hummel (2003). Graphic courtesy of Sarah Hummel.

This investigation suggests that Frisbee players should devote attention to improving their shoulder movement, in contrast to the often-emphasized wrist.

Figure 8.21. A player in the 2005 Kiwani Ultimate championships puts her shoulder into a forehand throw. Photo courtesy of Andrew Davis, www.freeheelimages.com.

Instrumented discs can also give some insight into the throw by documenting the acceleration and spin history of the disc (Figure 8.22). It may be possible to measure the disc deformation during a throw with strain gauges or other sensors.

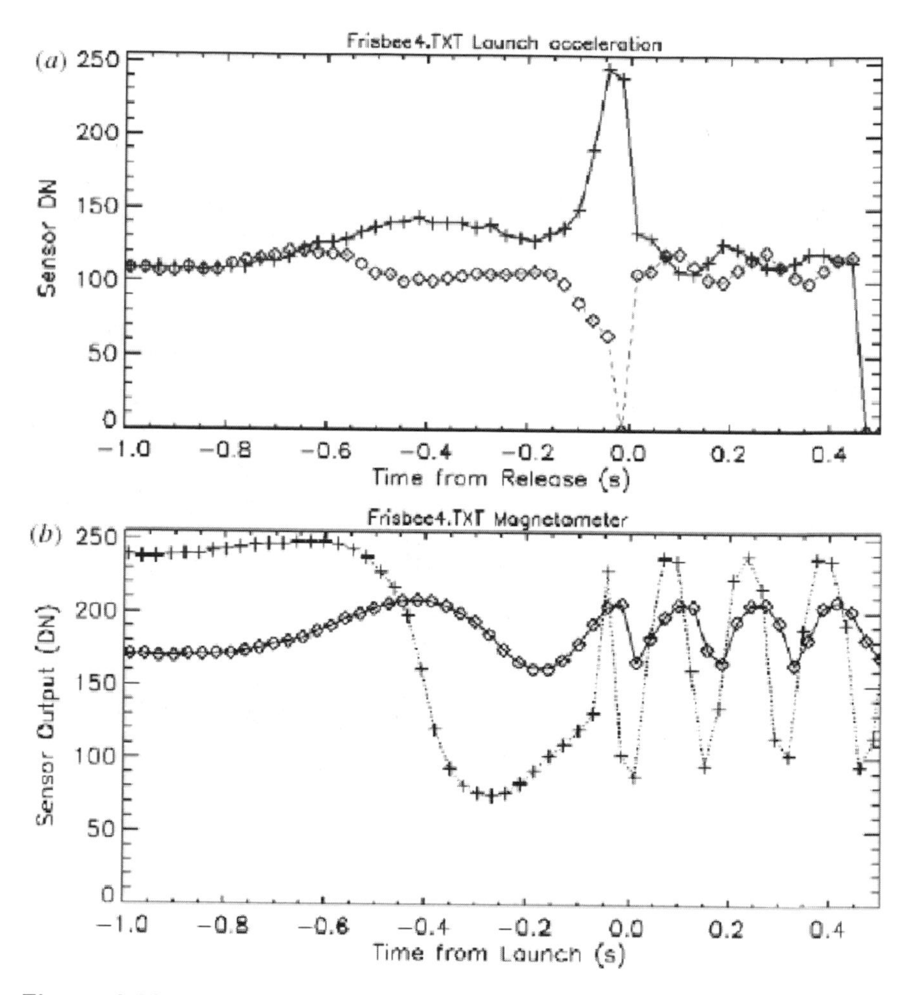

Figure 8.22. Accelerometers on a disc (a) show how about half of the total velocity acquired by the disc is picked up in a swing of about 0.4 s in duration, the rest being picked up in about 0.1 s by a snap of the wrist. The magnetometer record (b) shows how almost all of the spin is derived in this 0.1 second.

References

The 2003 paper by Potts and Crowther must be considered a benchmark work in the field. Hummel's Ph.D. thesis is also essential reading. Both items, and their websites, have lots of background and further references.

Higuchi, H., Goto, Y., Hiramoto, R., & Meisel, I., Rotating flying disks and formation of trailing vortices, *AIAA 2000—4001*, 18th AIAA Applied Aero. Conf., Denver, CO, Aug. 2000.

Hubbard, M., Hummel, S. A., 2000. Simulation of Frisbee flight. In *Proceedings of the 5th Conference on Mathematics and Computers in Sports*, University of Technology, Sydney, Australia. (at Hummel's website below)

Hummel, S., *Frisbee Flight Simulation and Throw Biomechanics*, M.Sc. thesis, UC Davis, 2003.

Johnson, S., Frisbee—*A Practitioner's Manual and Definitive Treatise*, Workman Publishing, 1975.

Lorenz, R. D., Flight and attitude dynamics of an instrumented Frisbee, *Measurement Science and Technology* 16, 738—748, 2005.

Lorenz, R. D., Flight of the Frisbee, *Engine*, April 2005b.

Lorenz, R. D., Flying saucers, *New Scientist*, 40—41, 19 June 2004.

Malafronte, V., *The Complete Book of Frisbee: The History of the Sport and the First Official Price Guide*, American Trends Publishing, 1998.

Nakamura Y. & Fukamachi N., Visualisation of flow past a Frisbee, *Fluid Dyn. Res.*, V7, pp. 31—35, 1991.

Potts, J. R., & Crowther, W. J., Visualisation of the flow over a disc-wing. *Proc. of the Ninth International Symposium on Flow Visualization*, Edinburgh, Scotland, UK, Aug. 2000.

Potts, J. R., Crowther, W. J., 2002. Frisbee aerodynamics AIAA paper 2002—3150. In *Proceedings of the 20th AIAA Applied Aerodynamics Conference*, St. Louis, Missouri.

Rohde, A. A., *Computational Study of Flow around a Rotating Disc in Flight*, Aerospace Engineering Ph.D. dissertation, Florida Institute of Technology, Melbourne, Florida, December 2000.

Schuurmans, M., Flight of the Frisbee, *New Scientist*, July 28, 127 (1727) (1990), 37—40.

Stilley, G. D., & Carstens, D. L., adaptation of Frisbee flight principle to delivery of special ordnance, *AIAA 72—982, AIAA 2nd Atmospheric Flight Mechanics Conference*, Palo Alto, California, USA, Sept. 1972.

Yasuda, K., Flight- and aerodynamic characteristics of a flying disk, *Japanese Soc. Aero. Space Sci.*, Vol. 47, No. 547, pp. 16—22, 1999 (in Japanese).

Frisbee Sports:
www.discgolfassoc.com/history.html
www.freestyle.org

www.whatisultimate.com/
www.ultimatehandbook.com/Webpages/History/histdisc.html

Frisbee Dynamics:
www.disc-wing.com
www.lpl.arizona.edu/~rlorenz
mae.engr.ucdavis.edu/~biosport/frisbee/frisbee.html

9
Spinning Cylinders and Rings

Although the modern Frisbee is perhaps the most familiar and popular flying spinning disc, there are a range of variants on the theme.

Discus

The sport of throwing the discus of course dates back to the ancient Greeks and the original Olympic games. An icon of athletics is Myron's famous statue (5th century B.C.) of Discobolus, the focused and muscular discus thrower. According to Greek legend, Apollo fell in love with a Spartan prince, Hyacinthus, but when they were practicing discus-throwing, the jealous god of the west wind, Zyphyrus, blew an ill wind and caused the discus to veer and strike a fatal blow to Hyacinthus.

Discuses were originally made of stone, then later bronze, with diameters of 17 to 32 cm and masses of 1.3–6.6 kg. The most common modern specification is for 2 kg and 22 cm diameter. The discus shape may be thought of as two flat cones stuck back to back; in other words the cross-section of a discus might be thought of as a rhombus, with the corners flattened and rounded. They are usually made of wood, with a steel reinforcement around the rim to prevent impact damage.

To first order, a discus throw is rather similar to a shot put. Since the discus is a somewhat heavy object (rather more massive than the forearm), the thrower employs as much whole-body and shoulder rotation as possible. A discus is somewhat larger and lighter than the shot, and thus experiences a slightly nonballistic trajectory. This of course must have been known to the Greeks, as otherwise the trajectory of Apollo's discus could not have been affected by wind! The development of lift and drag, which are a function of the angle of attack of the discus, makes the length of the discus throw dependent (unlike for a shot put) on the orientation of the discus when it is launched, so precise control is perhaps more important for discus throwing. Like a Frisbee, although to a lesser extent, a discus is thrown with a slight spin to give it some gyroscopic stiffness and thus keep the attitude somewhat constant.

Aerodynamics of the Discus

Frohlich (1981) has made numerical simulations of discus throws, using lift and drag coefficients determined from wind tunnel tests. The discus acts as a wing with a lift curve slope of about 0.03 per degree and a stall at 30 degrees or so at which lift drops and drag rises substantially. The pitch moment coefficient is roughly proportional to the lift coefficient.

The pitch moment does tend to cause a slight roll of the rotating discus in flight—deviations of perhaps 10 degrees can be observed. This effect is generally much less pronounced than on a Frisbee because of the much higher moment of inertia of the discus.

One result of these simulations is that a discus can in many circumstances be thrown further upwind than downwind. This rather paradoxical result is derived as follows. Upwind, the relative airspeed of the discus is increased relative to a no-wind case. The discus therefore develops both more lift and more drag. The lift has the effect of prolonging the flight. This flight extension provides a larger positive increment in the horizontal distance traveled than the negative increment due to the augmented drag.

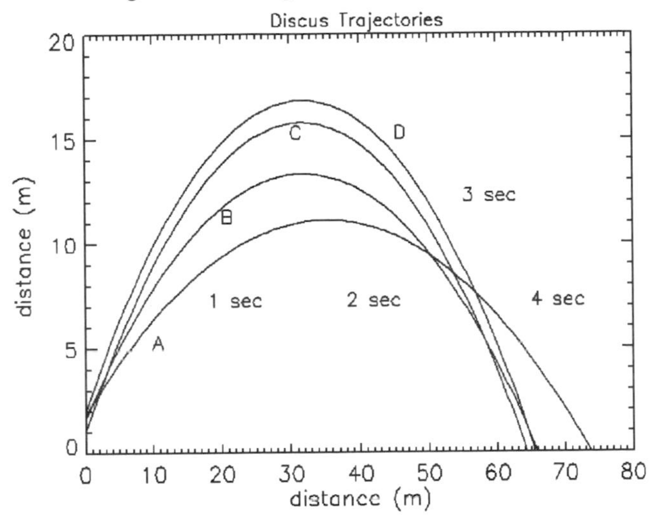

Figure 9.1. Trajectories of a 2 kg discus launched at 25 m/s at its optimum angle. The optimum angle is different for the different cases shown: (A) purely ballistic trajectory in vacuum, (B) in sea-level air with a 10 m/s tailwind (i.e., thrown downwind), (C) in air with no wind, (D) in air with a 10 m/s headwind. Even though (D) is thrown upwind, it goes further than the downwind throw.

Clearly, if the headwind were absurdly strong, the discus would be stopped in midair and would start flying backwards towards the thrower. Thus there is an optimum headwind speed for a given launch speed (25 m/s). There is a worst-case tailwind speed, at which the range achieved is a minimum. For modest tailwinds, the relative airspeed is progressively reduced and so the lift is compromised. On the other hand, for the highest tailwinds, the discus is carried downwind for a long throw. In between, at a windspeed of about 7.5 m/s, the range is a minimum of ~64 m, a meter or two shorter than the zero-wind case, and some 6 m or so shorter than for a 7.5 m/s headwind.

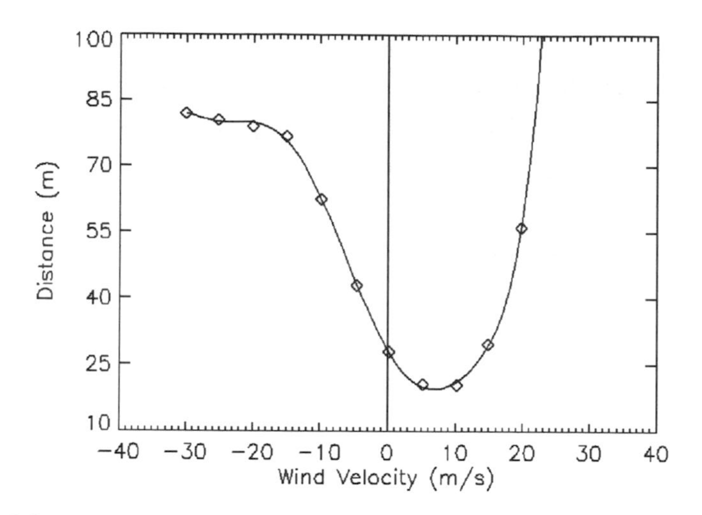

Figure 9.2. Flight length of a discus thrown with a speed of 25 m/s as a function of windspeed. The discus is thrown at its optimum angle in all cases. Negative winds correspond to upwind throws, and for modest windspeeds, allow the discus to go further than downwind throws.

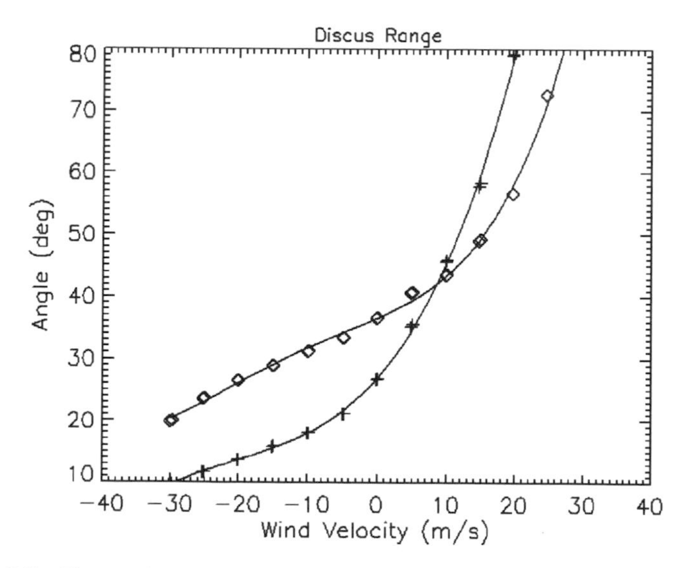

Figure 9.3. The optimum angles for discus throws. The plus signs indicate the angle that the discus velocity should make with the horizontal (the flight path angle) while the diamonds indicate the angle the discus plane should make with the horizontal (the attitude).

The effects of aerodynamic lift and drag are enough to affect sports performance at different locations—a throw at the same angle and velocity will travel 0.36 m less far in Mexico City (40°C, 2239 m) than on a cold day in Moscow (10°C, 120 m altitude); the air density is 30% greater in the latter case. Technically, there is also a small difference due to the greater acceleration due to gravity at higher latitudes on the rotating Earth—over the entire surface of Earth this variation is about 0.5%, leading to an increase in flight length of 0.34 m between equator and poles.

Table 9.1. Size and mass of flying discs. A clay pigeon is intermediate in ballistic coefficient between a discus and a Frisbee.

	Mass (kg)	Diameter (mm)	Mass/Area g/cm^2	Thickness (mm)
Men's Discus	2.0	221	5.2	46
Women's Discus	1.0	182	3.8	39
Wham-O regular Frisbee	0.087	227	0.21	31
Wham-O Ultimate Frisbee	0.175	280	0.28	31
Clay Pigeon	0.105	110	1.05	25

Clay Pigeons, Skeet

Another spinning object is the clay pigeon. Late in the 1900s, the substitution of clay targets for living birds in shotgun shooting was introduced and has become a recognized sport. (An alternative name for essentially the same sport, although various rules exist, is "skeet." Skeet is from a Swedish word for "shoot.") One or more clay discs are mechanically launched on flights lasting just a few seconds, flying typically 50 m or more. The shooter must follow the trajectory of the skeet and aim slightly ahead of it, since the cloud of shotgun pellets will take some hundredths of a second to reach the target, during which time the skeet will have moved.

Discs must be made of a biodegradable clay, be strong enough to sustain launch yet shatter satisfyingly when hit by shotgun pellets, be visible under various lighting conditions, and be inexpensive. Typical flights are 50–100 m in length. One standard clay pigeon (that defined by the Scottish Clay Target Association) is 110 mm in diameter, about 25 mm deep and weighs 105 grams. In fact, the "clay" is often chalk dust bound with bitumen (pitch) or wax.

The discs are launched by a catapult (or "trap"—originally real birds were released from traps) with a spring-loaded or motorized arm that flings the disc, usually along a plate or a track. As the rim of the disc rolls along the arm, it is given some spin to stabilize its flight (although it is noted that a wet arm, with low friction, will produce less spin). The spin can be important in the scoring of clay pigeon shooting, as well as in the dynamics of the flight. A weak crack formed in the disc by a slow or small pellet may not break apart under aerodynamic loads alone. However, the rapid spin can cause centrifugal fissioning of the disc.

As far as I can determine, the aerodynamic behavior of clay pigeons, and the evolution and effects of their spin, has not been scientifically documented, although the work of Stilley and Carstens (cited in the previous chapter) does come close. Another relevant study is that of Zdravkovich et al. (1998).

Other Flying Discs

There is a bewildering array of flying discs, and in particular the popularity of disc golf has introduced a large range of golf discs with different weights and shapes. Golf discs are usually smaller than a Frisbee, but with a thicker rim, and a more ellipsoidal than flat upper surface. Weights of 150–180 g and diameters of 20 cm are typical. Some manufacturers, such as Innova, Inc., parameterize the properties of their range of literally dozens of discs with "speed," "glide," "turn," and "fade," rated with the integers 1–9, so that like a conventional golfer choosing a "5-iron," a disc golfer can choose the best disc for a given shot. The first three of these properties presumably relate directly to drag, lift, and low-α pitch moment coefficients (scaled by mass and moment of inertia). "Fade" refers to the amount of left turn at the end of the flight—this will relate to the pitch moment at high angle of attack as discussed in the previous chapter.

An interesting golf disc by Aerobie (see below) is the "Epic," which has a conventional-looking upper surface. However, the cylindrical cavity inside the disc is smaller than most discs, and is offset from the center. The offset permits a suitably narrow region for gripping the disc, but the offset displaces the disc center of mass further from the fingers, and in effect lengthens the arm of the thrower, allowing for a faster launch. Whether there is also a significant aerodynamic or gyrodynamic advantage remains to be seen.

Among the many Frisbee variants, a product worth discussing is Aerobie's Superdisk, a disc with a rather flat spoiler rim (made of a comfortable rubber). This disc is allegedly easier to throw than a conventional Frisbee, but does not go as far. If the claim of easy throwing is true, it is presumably a result of the spoiler aerofoil having a pitch moment coefficient that is small over a wider range of angle of attack. It would be interesting indeed to see wind tunnel measurements of the aerodynamic coefficients of this disc.

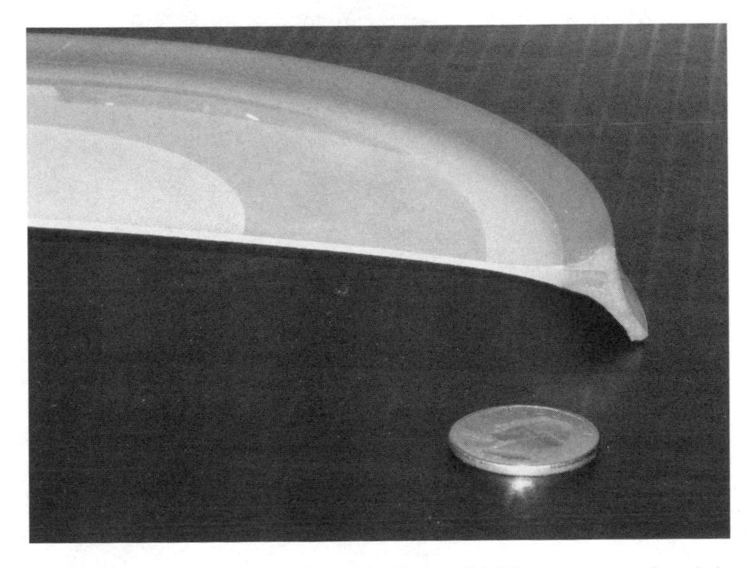

Figure 9.4. Cross-section of an Aerobie Superdisk. The upper surface is less cambered than a conventional Frisbee, but the rim is quite flat and has a spoiler on the upper surface.

Spinning Rings

The conventional Frisbee of course owes its origins to pie tins or lids being thrown. These evolved into an object that could be cheaply mass-produced and that had a set of flight characteristics that lent themselves to recreational use, generating a variety of flight profiles and making games such as Ultimate Frisbee possible. However, because of the thickness of the disc required to suppress the pitch moment, the draggy Frisbee does not permit flights of extreme length. The throwing toy that achieves this goal has a rather different shape, and was developed by Alan Adler, founder of Aerobie, Inc., (formerly Superflight, Inc.) in Palo Alto, California, in the 1970s.

Figure 9.5. Alan Adler, with the ring/boomerang prototype that eventually led to the Aerobie flying ring.

Adler, originally an electronics engineer, but an all-round tinkerer, was in fact experimenting with candidate boomerang designs, and after one design showed good flight characteristics but no hope of returning (it was a ring with two arms—see Figure 9.5), he decided to explore nonreturning designs and eventually refined an object that permits very long-range throws. (In fact, while millions of these flying rings have been sold, they typically fly too well—or too fast—for short-range recreation, and do not exhibit the range of different flight profiles as do conventional Frisbees.)

Although Adler uses computer simulations to determine post hoc the aerodynamic performance of various designs, the design process itself does not use wind tunnel or computational fluid dynamics simulations, but is rather one of reasoned design followed by field trials. A team of test throwers will try out various handmade designs before settling on the optimum, after which a mould is then fabricated for production.

Chakram

The flying ring was of course not invented in California in the 1970s. Flying rings have a much longer history elsewhere, most notably in India. The chakram (also chakra, chakrum, chakar, etc., and sometimes a "war quoit") is a throwing ring weapon, although perhaps more ceremonial than militarily significant. They are widely reported in legends such as the Mahabharata—while Krishna liked the battle-axe, the god Vishnu favored the chakram. Between the 16th and 19th centuries Sikh soldiers used them against the Moghuls, with infantry throwing them in volley fire at ranges of a few tens of meters. It was reportedly used by street thugs in 1930s Calcutta. Most recently, these weapons were made popular by the TV fantasy series *Xena: Warrior Princess*, whose protagonist rather improbably flings them around. The ricochet throws portrayed in the series are particularly nonphysical.

The chakram exploits the stability of a spinning ring with the aerodynamic performance of a flat plate. It should be noted that for the typical scale of thrown objects in air ($V \sim 20\,\mathrm{ms}^{-1}$, $\partial \sim 20\,\mathrm{cm}$) the flight Reynolds number is rather low—around 20,000. In this regime, thin plates (ideally cambered, but even flat) perform rather better than thick airfoils.

Figure 9.6. An illustration from Egertons book showing how the ring is thrown by the Sikhs, by twirling around the fingers. Ornate patterns are usually engraved on the ring.

Usually chakrams are around 20 cm in diameter, with a chord of 2 cm or more. The outer edge is ground to form a sharp blade, and the object is usually spun and thrown by twirling, hula-hoop-style, the rounded inner edge around the finger (although they are sometimes thrown in an underarm Frisbee-style, with the ring gripped between finger and thumb). Although relatively small in area compared with Frisbees, being made of steel or brass with densities of ~8000 kgm^{-3} (vs. around 900 kgm^{-3} for polypropylene), these objects are still quite massive (~200 g), and thus have high wing loading (~10^4 kgm^{-2}). Such devices are not toys—in the right hands they are able to cut bamboo poles or slice watermelons at ranges of tens of meters.

Some chakrams are simply flat aerofoils, although others are formed as if the ring were on the surface of a sphere or cone—the aerodynamic implications of this are discussed in the following section. In general they are plain, although some intricately engraved examples exist (it seems doubtful that engraving modifies the aerodynamic performance via surface roughness effects, however).

Recreational Flying Ring Design and History

The flying ring, most commonly encountered in modern times as the Aerobie, is an attempt to circumvent the flying disc's most salient problem, namely the forward center lift and its resultant pitch-up moment. Almost all aerofoil sections have their center of pressure at the quarter chord point, while the center of mass is at the half chord.

A ring-wing gets around this problem in part by pure geometry: it can be considered by crude longitudinal section as two separated wings. While the lift on each wing will act forward of the center of each, if the two wings have a sufficiently short chord, this lift offset will be small compared with the overall diameter of the vehicle.

In reality, for recreational applications at least, the diameter of the vehicle must relate somewhat to the size of the human hand; portability constraints argue against large diameters. Furthermore, the requirement of low drag will require a thin wing section, which makes it difficult to provide torsional stiffness for large diameters. One could in principle make a large diameter, high–aspect ratio ring-wing with excellent aerodynamic performance, made from a stiff metal. This, however, might perform better as a weapon than as a toy!

A modest diameter, even 30 cm or so, introduces two new complications to our 2-dimensional idealization. First, the large chord makes the trailing wing shorter than the leading wing.

The second, and more important, issue is that the trailing wing is immersed in the downwash from the leading wing. This has the effect of reducing its effective angle of attack and throwing the ring out of balance by reducing the lift on the trailing wing.

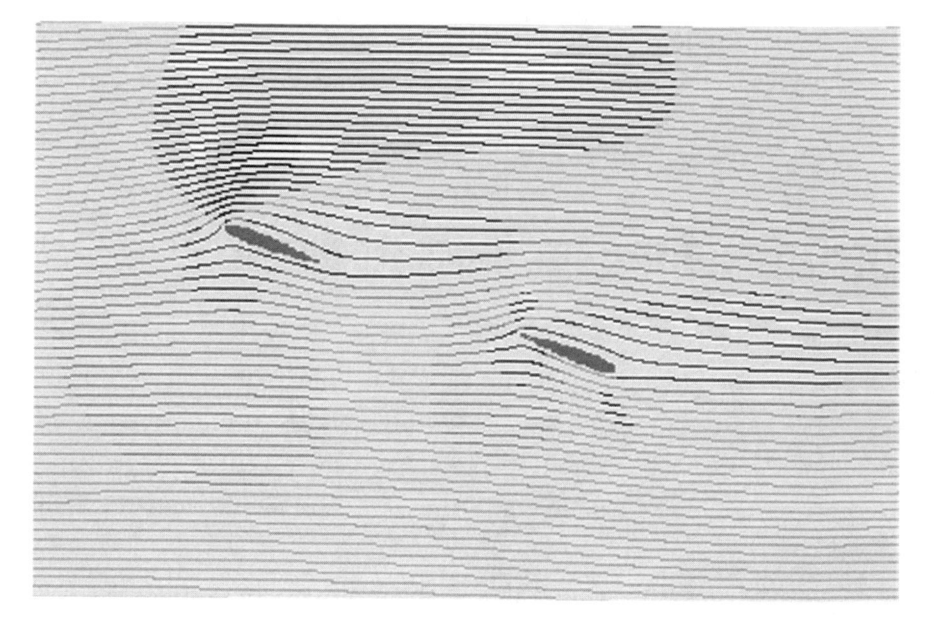

Figure 9.7. Downwash. Two NACA-0012 aerofoils are flying back to back right to left in this CFD simulation showing the streamlines. It can be seen that the flowfields of the two aerofoils interact—the right foil is immersed in the flow directed downwards by the left one. The shade of the streamlines indicates the local pressure in the flowfield.

One approach that was tried initially in the "Skyro" (the first flying ring sold by Aerobie—it sold around a million in the 1970s) was to use a rather symmetric aerofoil, but to have it angled such that the wing formed a cone.

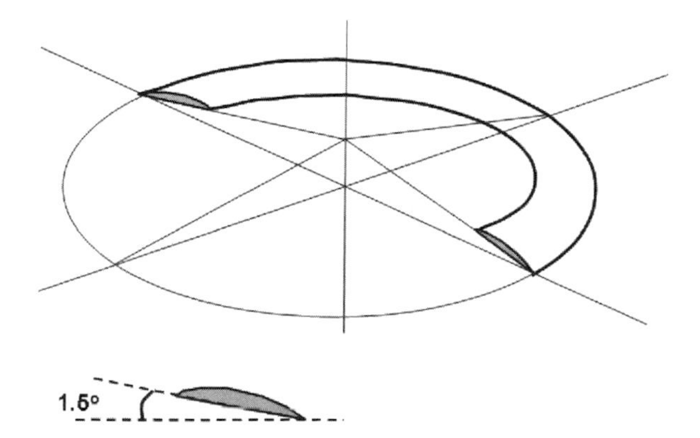

Figure 9.8. Drawing showing the conical layout of the Skyro, forcing the trailing wing to have an adequate angle of attack in spite of the downwash from the leading wing.

The trailing wing therefore was mounted at a higher angle of attack to the freestream flow than was the leading wing, and thus when downwash was taken into account the two were at a comparable angle of attack. The two wings thence had the same lift coefficient and the ring flew in a trimmed condition. However, this tuning (a cone angle of only about 1.5 degrees was necessary) was only strictly correct at one flight speed, and thus a perfectly trimmed condition was only found during a portion of a typical flight. That said, the conical design was a significant improvement.

Tuning the vehicle over a range of flight conditions instead needed a carefully selected aerofoil section, which had a lift curve slope higher for outwards (trailing) flow than for inwards (leading wing) flow. The higher lift curve slope therefore compensated for the lower angle of attack, such that the resultant lift coefficients were similar. The aerofoil with this characteristic had a rather severe reflex, almost as if it had to spoilers on its trailing edge.

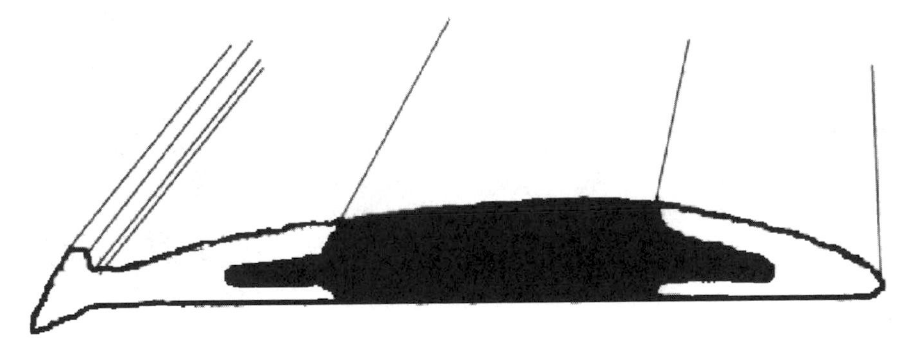

Figure 9.9. The rather special aerofoil used in the Aerobie. Note, counterintuitively, that the outer edge of the ring is to the left—it almost looks as if the leading edge of the ring flies "backwards."

Some dye-flow experiments have been conducted at NASA's Dryden Research Center. These show very different characteristics for the forward and reverse flow, as might be expected for lift coefficients that vary by a factor of 2. The tests show that the flow is quite turbulent with strong separation from the spoiler when it is "forward," as might be expected.

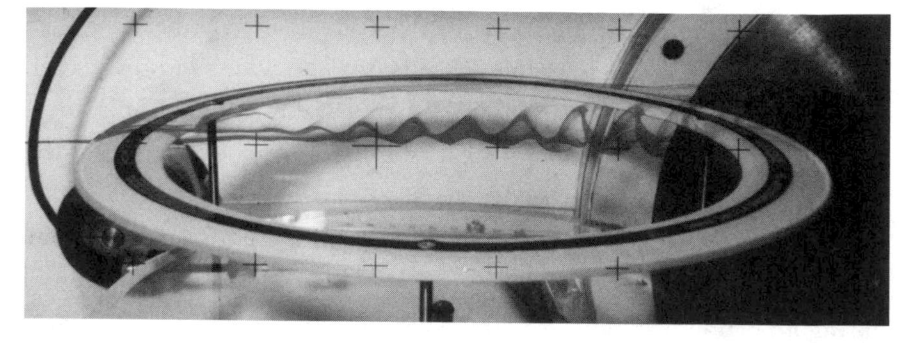

Figure 9.10. Dye-flow experiment in a water tunnel over an Aerobie. Note how the trailing side (right) is immersed in the oscillatory wake from the leading side. NASA Dryden Photo EC91 120-4, courtesy of Alan Adler.

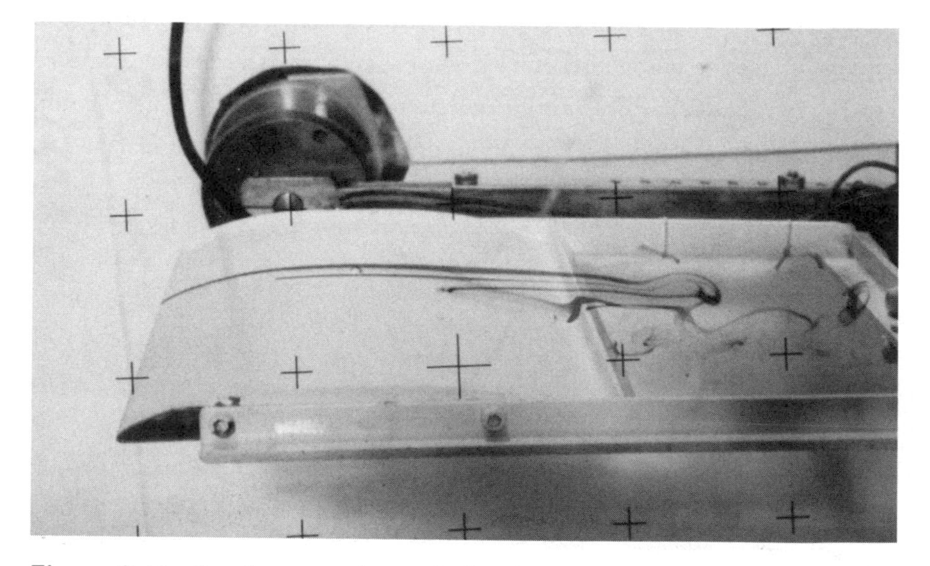

Figure 9.11. Dye-flow experiments in Dryden water tunnel, this time over a 2-D trailing side wing section, showing flow separation. Again, flow is from left to right. Notice the motor used to vary the angle of attack (this run at 2°) and the cross-shaped reseau marks to act as fiducials for distance measurement. NASA Dryden Photo EC91 120-1, courtesy of Alan Adler.

Even so, the flying ring is rather thin (~3 mm) compared with a Frisbee, and thus has much lower drag. As a result, the range achievable with a flying ring is much further—the present record is some 1400 ft. This flight was made by throwing along a ridge (so it may have gained from some updraft lift) although it was terminated prematurely, ending about 1.5 m above the ground by striking a bush. This flight lasted only 7 seconds—much shorter than many boomerang flights; although the lift:drag ratio is very good, the actual lift coefficient is small and thus the flight speed must be fairly high.

As with boomerangs and Frisbees, material selection is important. The ring must be adequately weighted to efficiently extract energy from the throw, and to provide sufficient moment of inertia to remain spinning. An additional consideration in this sort of application is compliance, as a metal Aerobie would be rather unpleasant to catch.

The flying ring (and its boomerang counterpart) are constructed with a polycarbonate "backbone" which is placed in a mould into which a lower-density rubber is injected. This combination yields the desired density, as well as the desired compliance and "memory" (the ring can be "tuned" slightly by flexing it—were it perfectly elastic, such adjustments would be impossible).

Figure 9.12. Cross-section of the Skyro flying ring, a predecessor of the modern Aerobie. Notice the conventional wing section, and the construction from two plastics—a central stiffer skeleton with an outer more flexible material for comfortable catching. Photo by the author.

In fact, a quite reasonable flying ring for indoor use can be constructed in a matter of minutes from corrugated cardboard. A ring can be marked out and cut. The face-sheets of the cardboard can be delaminated a few mm from the outer edge on both sides to give the reflex (stubbing the cardboard with a fingernail will do the job quite well). This ring will fly very straight, although collisions with walls and furniture mean it needs to be retuned periodically.

Ring Wings and the X-zylo

A wing can be made into a ring by making a circle in another dimension, with the wing chord parallel to the axis through the center of the circle, rather than orthogonal to it. The wing thus forms a cylinder through which the air flows as it flies through the air.

A completely symmetrical circular wing will develop no lift at zero angle of attack. However, with a positive angle of attack, the wing will direct air downwards, and therefore generate lift. Although such a design might at first seem rather absurd (and wasteful, in that part of the span is oriented vertically, so that it doesn't contribute to lift), there are two advantages of this configuration. First, the wing has no ends, and so there is not the same shedding of tip vortices that leads to induced drag. Second, there are structural advantages, as a cylinder can be made strong and light. By connecting the wing into a circle, load paths can be shorter and can be shared.

Although some fanciful artists, impressions have been sketched of ring-wing airliners, one has to wonder whether this would pass the "laugh test" of prospective passengers. An aircraft must not only fly, but look like it is able to. However, ring-wing architectures are likely to find application in unmanned aerial vehicles. A vehicle can use a fan inside the ring for vertical take-off (i.e., a large ducted fan), with the duct offering noise reduction and containment of the rotor, an important factor in operations in tight spaces where personnel may be nearby. For longer traverses, the vehicle can transition by tilting forward and accelerating, to derive most of its lift more efficiently from the ring-wing rather than the rotor.

The other application is of course as a toy. The ring-wing (sometimes called a "bishop's hat") is a surprisingly effective configuration for an airplane made from a single sheet of paper. Indeed, it offers substantial economy in fabrication time over conventional designs, in that only 3 folds (plus one tuck) are needed, as against 5 or 7 for even a simple conventional paper dart. The ring-wing shape thus formed has a natural tendency to roll its long side downwards, and is therefore aerodynamically stable. It is challenging to trim these ring-wing planes, with the result that they often tend to porpoise, accelerating into dives followed by pull-up and stall, then repeating the cycle. (It is also possible to make paper airplanes with a conventional or canard ring-wing configuration—a main ring-wing, with a stabilizing foreplane ring-wing. However, although aesthetic, these are hardly easy to construct.)

The other approach to stabilization is to spin. This requires radial symmetry, such that the mitre shape of the paper bishop's hat will not work—the wing must instead be simply a cylinder. This is the configuration adopted by the X-zyLo toy.

The X-zyLo is a 25 g toy about 96 mm in diameter and about 60 mm tall. The leading edge is weighted by thickening the plastic, while the trailing edge has a wavy pattern. Remarkably, for an item so light (six times lighter than a typical Frisbee or baseball), an X-zyLo has been thrown some 200 m.

The instructions say to throw "like a football" (i.e., an American football), projecting it forward with a clockwise spin as seen from the thrower. The X-zyLo tends to sweep towards the left, suggesting it has a slight pitch-down tendency in flight, which is gyroscopically modified into a yaw to port. It is not clear if the scalloped trailing edge has an aerodynamic function or is purely decorative.

The X-zyLo manufacturing and marketing operation (the William Mark Corporation) was set up by a laid-off California aerospace worker, William Forti. The design actually came from his son, Mark (a marketing student at Baylor University at the time), who was experimenting with paper airplanes when he was surprised at how well a ring flew. They patented the design and began marketing the plastic version (the X-zyLo Ultra—a glow-in-the-dark version is also available). The company's literature claims that NASA experts were unable to explain why the product flew so well. I suspect, however, that at least *some* NASA engineers might not be surprised, since the principles of aerodynamics and gyroscopic motion are not *that* arcane. Nonetheless, the flying performance of such a light and unconventional-looking object is quite remarkable.

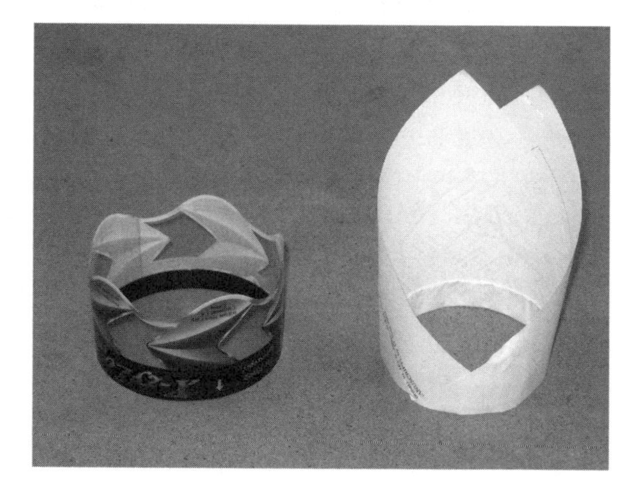

Figure 9.13. The 25 g spin-stabilized X-zyLo toy that can be thrown some 200 m. The nonspinning bishop's hat paper ring-wing offers less performance, but is very quick to make.

It is easy to make a paper version simply by folding over a sheet to make a weighted leading edge, and using tape or a staple to seal the ring in place (friction between the paper is sometimes enough on the asymmetric bishop's hat, although it can be augmented with a fold or three to secure the paper in place; this doesn't work on a pure ring.) It can be readily found by experiment that stability is poor for aspect ratios greater than about 1.

References

Adler, A., The Evolution & Aerodynamics of the Aerobie Flying Ring, note available at www.aerobie.com

Egerton, (The Honourable Wilbraham, M.A., M.P.), *An Illustrated Handbook of Indian Arms*, 1880.

Frohlich, C., Aerodynamic effects on discus flight, *Am. J. Phys.* 49(12), 1125–1132, 1981.

Zdravkovich, M., A. J. Flaherty, M. G. Pahle, and I. A. Skelhorne, Some aerodynamic aspects of coin-like cylinders, *J. Fluid Mechanics*, 360, 73–84, 1998.

www.aerobie.com
www.discwing.com
http://flyingproducts.com/
www.innovadiscs.com
www.xzylo.com

10
Spinning Aircraft and Nonspinning Disc Aircraft

Although the central theme of this book is spinning flight, mention deserves to be made of more or less conventional (piloted) aircraft which have had substantially disc-shaped planforms. In this chapter we also describe briefly a number of aircraft that use conventional wings but also carry large discs. In several cases these radomes rotate, qualifying them as "spinning discs," although we shall see that the spin in these cases is not aerodynamically significant.

Further, although the scope of this book does not extend to conventional helicopters (the aerodynamics and gyrodynamics of which are indeed interesting but whose treatment needs a whole other book), I also add to the miscellany in this chapter a couple of spinning rotor vehicles (one a UAV, one a toy) which are distinct from helicopters in that the whole vehicle spins.

Finally, no discussion of these topics would be complete without mention of the closest thing to a flying saucer built on Earth, the Canadian Avrocar.

Nonspinning Disc-Winged Aircraft

A circular planform of course has by definition an aspect ratio of 1. This is in stark contrast to the high aspect ratios that are generally demanded for subsonic aircraft due to the low induced drag ("vortex drag") that results. For example, high-performance sailplanes may have extremely long and narrow wings with aspect ratios in excess of 20 to maximize their glide performance. Thus in a conventional aerodynamics sense, disc-wings are poor performers. However, at slow speeds and thus high angles of attack, the strong vortex lift from low-aspect wings like disc-wings and delta-wings can compensate for the vortex drag that would otherwise compromise their cruise performance.

Figure 10.1. The strong wingtip vortex as air spills from under the wing visualized by smoke in a test by NASA Langley at Wallops Island. NASA Image EL-1996-00130.

Aviation has always demanded performance from the structural engineer; weight in an aircraft is at a premium. The symmetry and intrinsic strength of circular shapes has therefore had a recurrent appeal.

In the earliest days of aviation, before the aerodynamic advantages of high aspect ratio were appreciated, many early attempts at aircraft had vaguely disclike planforms. Additionally, in recent years, interest in Unmanned Aerial Vehicles (UAVs) has grown, and in micro-UAVs in particular, since advances in electronics now permit video transmission from camera systems weighing only a few tens of grams. At the low Reynolds numbers associated with these 30 cm scale vehicles, a low–aspect ratio wing performs quite well. Furthermore, the aerodynamic elegance of slender wings amounts to nothing if they break off in a soldier's backpack; low–aspect ratio wings offer structural and packaging advantages.

Lee–Richards Circular Airplane

A remarkable early aircraft was developed in northern England by Cedric Lee in 1911–1912: a glider (and later a powered version) with an annular wing planform. The wing was circular, with a circular hole of a half-diameter in which the pilot sat. In some versions the aircraft was in fact a biplane, with an upper wing the same as the main wing, or in some cases just the front half of the main wing. Control was by more or less conventional elevators, which were also used for roll control as ailerons, and a tailplane. The 7 m span craft, with pilot and ballast, weighed some 300 kg.

The plane was launched from a track, being pulled by a rope attached to a weight that dropped from a tripod. The aircraft performed rather well, with a glide ratio of up to 8 : 1, and could be flown safely to its stall point with an angle of attack of 30 degrees. Beyond the stall, the aircraft parachuted down. This forgiving characteristic was by no means typical for early aircraft.

Figure 10.2. Lee–Richards powered biplane, with two annular wings and a conventional tail. From lantern slide in the collection of the Royal Aeronautical Society.

Arthur Sack AS 6

As with so many exotic aircraft types, engineers in World War II Germany appear to have tried the disc-wing concept (other innovations include the jet-powered Me 262, the rocket-powered Me 163, the Horton Flying Wing, etc.).

After several model experiments demonstrated the concept, Arthur Sack (a Bavarian farmer and plane modeler) built a small test aircraft in 1944 from plywood and scavenged parts (including the cockpit from a wrecked Messerschmidt-109). These tests were not spectacular successes, but did help reveal some of the intrinsic problems of this type of aircraft. The first difficulty encountered was that the control surfaces were in the wake of the wing while taxiing.

One disadvantage of the narrow planform is the short moment arm for roll control: on the AS 6's fourth test (really no more than a hop) the torque from the propeller caused the aircraft to bank to the left. A further similar test flight was later made by ar Me 163 pilot (pilots of this aircraft might already be considered suicidal!), resulting in collapsed landing gear. The prototype, nicknamed the "Flying Beermat," was thereafter scrapped.

Vought's "Flying Pancakes"

Charles Zimmerman was an engineer working for NASA's predecessor, NACA. An early contribution of his was the investigation of aircraft spin characteristics in a dedicated spin wind tunnel.

Figure 10.3. The 15-foot spin tunnel at Langley in 1935. NASA Image EL-2001-00112.

Figure 10.4. A model of the XB-47 free-flying in a larger spin tunnel at Langley in 1945. NASA Image EL-2000-00235.

He explored the aerodynamic properties of low–aspect ratio wings, and became convinced that they offered useful performance. He also had the interesting idea that by mounting the propellers near the wingtips (not a major structural penalty given the modest span), the propwash would tend to cancel out the tip vortices, giving much better induced drag performance than the low aspect ratio would suggest. His initial idea was entered in a NACA design competition in the hope it might be capable of hover like a helicopter, but with high-speed cruise. The idea was rejected as being "too advanced" (Ginter, 1992).

Later, now working for the Vought Corporation, he sought to design an aircraft that would have a low airspeed for operation from carriers. After promising model tests (one result of which was the addition of horizontal stabilizers—i.e., a tailplane) a full-scale wood-and-fabric experimental version, the V-173 was first flown in 1942. The low-speed performance was good, since almost the entire lifting surface was immersed in the flow from the two large propellers. It was very stable, difficult to put into a spin, and difficult to stall. The aircraft had a 23 ft wingspan, and weighed about 3000 lbs.

It had a tall storklike undercarriage, which held it at some 22 degrees angle of attack to generate enough lift. The high-α flying meant the controls were mushy—the aircraft was slow to respond, a characteristic that often is concomitant with good stability. One aerodynamic feature that is interesting is that close to the ground, the plane tended to pitch down, and a special trailing edge flap was added to relieve this effect. Cockpit vibration was also a problem.

Based on the somewhat promising characteristics of the V-173, Zimmerman hoped a more powerful version—with the same wing span and area (427 ft^2) but four times heavier and with four times as much power—would be able to fly over the remarkable speed range of 40–425 mph (not quite the hover that had been hoped for). Note that this remarkable factor of 10 in speed does not quite imply a 100:1 variation in lift coefficient, in that the propulsive thrust balances part of the weight, and the meaning of lift coefficient in a wing immersed in propwash may not be quite clear (see also the Turboplan discussion later).

Prototypes of an operational-scale aircraft, the XF5U, were built in 1945, but flight-testing was delayed by the late delivery of special flapping propellers that would be needed to minimize the vibration due to asymmetric loads at high angles of attack. Other development problems included the formidable engine transmission system. By the time the aircraft was ready for flight trials, naval interest had shifted to jet-powered aircraft.

Figure 10.5. Vought V-173 in the wind tunnel at NASA (then, NACA) Langley Research Center. Note that this early variant had two-bladed propellers. Notice also the transparent canopy on the underside of the aircraft—since the plane took off and landed at high angles of attack, the ability to look down was essential. NASA image.

Figure 10.6. Vought V-173 in flight. Notice the three-bladed propellers and the tall undercarriage, both to hold the aircraft at high angle of attack for lift, and to keep the propeller tips from striking the ground! Photo courtesy of Vought Aircraft Industries, Inc.

Figure 10.7. The prototype Vought XF5U-1. The four-bladed propellers and the bulges between cockpit and propeller only give a hint of the 4-fold increase in engine power compared with the V-173. The prototype never flew, however. Note the footprints on the wing/body. Photo courtesy of Vought Aircraft Industries, Inc.

Aircraft with Circular Radomes

Moving ahead several decades to the age of jets and missiles and the resultant need for long-range radar surveillance, several aircraft have been developed with large rotating radar antennas faired in discus-like radomes mounted above the fuselage. These radomes are in essence disc-wings. Perhaps the best-known of these aircraft are the EC-2 Hawkeye and the E-3 Sentry.

The radome of the EC-2 Hawkeye is some 7.3 m in diameter, about one third of the wing-span. The area of the radome is significant—some 41 m^2, compared with the reference wing area of 65 m^2. The radome thus can easily generate its own weight as lift. The most significant aerodynamic perturbation is the disruption of the flow onto the tail due to the radome and its support (the radome can in fact be lowered slightly to

Figure 10.8. Hawkeye on the deck of an aircraft carrier. Note how its wings fold to save space in the cramped carrier. The folded wings accentuate the large area of the radome. US Navy photo.

facilitate accommodation in the cramped confines of an aircraft carrier). The aircraft therefore has a heavily modified tail, with two outboard tailfins.

The not inconsiderable aerodynamic and structural penalties of the radome do bring substantial capability. The Lockheed Martin AN/APS-145 radar is capable of tracking more than 2,000 targets, and is able to detect aircraft at ranges greater than 550 km. One radar sweep covers 6 million cubic miles. The twin-prop aircraft has a maximum endurance of about 6 hours, and a combat radius of 1500 km.

Although the radome rotates, it barely qualifies as a spin. Rotating at 5–6 rpm gives a circumferential speed of ~ 2 m/s; the aircraft's cruise speed is some 259 knots (480 m/s) and thus the advance ratio is ~ 0.005. Rotational effects on the flow such as the Robins–Magnus force are tiny and can therefore be neglected.

Figure 10.9. EC-2 Hawkeye arrives at NASA's Dryden Flight Research Center for load tests. NASA Photo EC04-0310-06 by Carla Thomas.

Figure 10.10. The aerodynamic loads on a radome are significant. Here technicians at NASA Dryden prepare for major structural load tests on an EC-2 Hawkeye. NASA Photo EC04-0360-50 by Tony Landis.

The much larger and faster E-3 Sentry (often referred to by the generic term Airborne Warning And Control System, AWACS) is a modified Boeing 707/320 commercial airframe. It has a similar rotating radome, 9.1 meters in diameter, 1.8 meters thick, which is held 3.3 meters above the fuselage by two struts. It also rotates once every 10 seconds. This $270 million aircraft can fly unrefueled for some 11 hours, and has a range of some 9200 km. It has a flight crew of four, plus between 13 and 19 mission crew to operate the radars and perform communications, etc.

An additional AWACS variant used in Japan is broadly similar in specification, but uses the airframe of the twin-engine 767 airliner. This platform allows more floorspace for the mission crew on the long surveillance missions.

Aircraft of roughly similar configuration have been designed in the former Soviet Union. The Beriev A-50 Mainstay (based on the Ilyushin Il-76 "Candid" transport aircraft) is a broad equivalent of the E-3, but its heavy radar gives shorter endurance. A medium-sized twin-jet aircraft, the An-71 Madcap, was also developed in the Ukraine (with a

Figure 10.11. The E-3 Sentry. NATO photo.

Figure 10.12. Head-on view of the E-3 with its massive radome. Note that the support structure uses two struts to provide adequate stiffness. This arrangement may also minimize wake effects on the tailfin. NATO photograph.

disc-shaped antenna installed, remarkably, on top of a forward-swept tailfin), but following the breakup of the Soviet Union was not adopted. The (Russian) Yakovlev Yak-44 was a twin-prop carrier-borne aircraft, rather similar to the E-2, initially selected in preference to the An-71, but shelved as the Russian carrier fleet withers away.

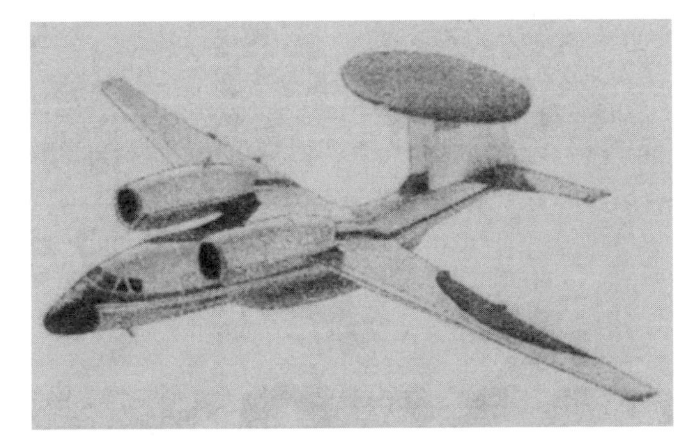

Figure 10.13. Sketch of the An-71 with its nonrotating radome mounted on the tail.

Disc-radome aircraft may be approaching obsolescence, in that the radome requirement derives from the need for large aperture (to achieve high directivity, and thus sensitivity) for the radar beam, while permitting 360 degree azimuth coverage, without rendering the aircraft unflyable. These requirements lead to a large, circularly symmetric rotating structure. However, modern signal-processing electronics permit the angular coverage to be met with a phased-array (i.e., electronically steerable) antenna. This must still be large to achieve the required sensitivity, but can be mounted conformally (or at least symmetrically) on the aircraft—the most recent early warning aircraft (the 737-derived Wedgetail for the Australian Air Force) has an antenna rather like a fish's dorsal fin.

The Avrocar

The most exotic disc-wing aircraft is the Avrocar, which almost perfectly resembles a flying saucer. The early (1952–1954) concepts by the Canadian division of A. V. Roe limited (makers of the Lancaster bomber, used to drop the Bouncing Bombs in chapter 13) was for a vertical take-off supersonic aircraft with a disc planform. Some of the very earliest tests, remarkably, were made with a small disc-wing mounted on a frame on the hood (bonnet) of a Pontiac sedan car. This was driven at up to 85 miles an hour!

The idea of a disc-wing became married with other concepts that held the promise of dramatic performance—since unlike on a conventional aircraft, the engine diameter could be very large, a radial engine geometry not unlike Briton Frank Whipple's earliest jet engine (rather than the more slender German Fritz Ohain axial-flow engine layout that became adopted worldwide) could be used. Also, flow ducted by exhaust gas from the engine might enhance lift, by exploiting the relatively newly known Coanda effect.

After some intermediate ideas supported by the U.S. Air Force, the project focused on a disc-shaped subsonic VTOL vehicle, capable of survival in a battlefield environment without the benefit of long airfields. Initial performance requirements for the Avrocar were a ten-minute hover capability in ground effect and 25 mile range with a 1000 lb payload.

A large (1.6 m) fan in the center of the 6 m diameter aircraft generated lift. The fan (called a turborotor) was driven by the ducted exhaust from three small turbojet engines. Fan air was ducted out through an annular nozzle at the base of the vehicle. There were two separate cockpits, on opposite sides of the disc (to effect balance—a single cockpit could not be mounted at the center since this was where the fan was).

A sophisticated control system was needed to respond to attitude perturbations (including the significant pitch-up moment) and control inputs, taking into account the gyroscopic action of the turborotor, the mounting of which permitted some movement relative to the vehicle as a whole.

Although initial expectations were high, the performance of the prototype vehicle was disappointing, largely due to pressure losses in the ducts and thrust degradation due to recirculation of exhaust air back into the turborotor. Specifically, the aircraft could not develop enough thrust to match its own weight, except when it was low enough to feel the "ground effect." Much a hovercraft, the cushion of air beneath the vehicle was retained by the presence of the ground. Above about 0.6 m, the cushion became unstable, and large oscillations in pitch and yaw ("hubcapping") developed which could not be arrested by control inputs. Modifications to the ducts for the downward jet were made which improved matters, but the vehicle still became unstable above horizontal speeds of about 50 m/s.

With control improvements, the vehicle was capable of steady, trimmed flight out of ground effect, but high angles of attack and high forward speed (180 m/s) were needed to do so, and pitch stability remained problematic. Experiments with a large tailplane were unsuccessful, presumably because the tail was in the downwash of the disc as a whole. An interesting problem was the asymmetry in yawing performance—a 90-degree turn to the right took 11 seconds, over twice as long as for a 90-degree turn to the left.

The enduring control problems led to cancellation of the program, which may simply have tried too many radical innovations at once. The strong influence of ground effect in early tests may have made the performance predictions overly optimistic—if the vehicle had been conceived as a sort of hovercraft that could make short hops in the air, the project might have proceeded in a more robust fashion. The vehicle (of which two prototypes were built, both now in U.S. museums) will endure as a magnificently bold departure from convention.

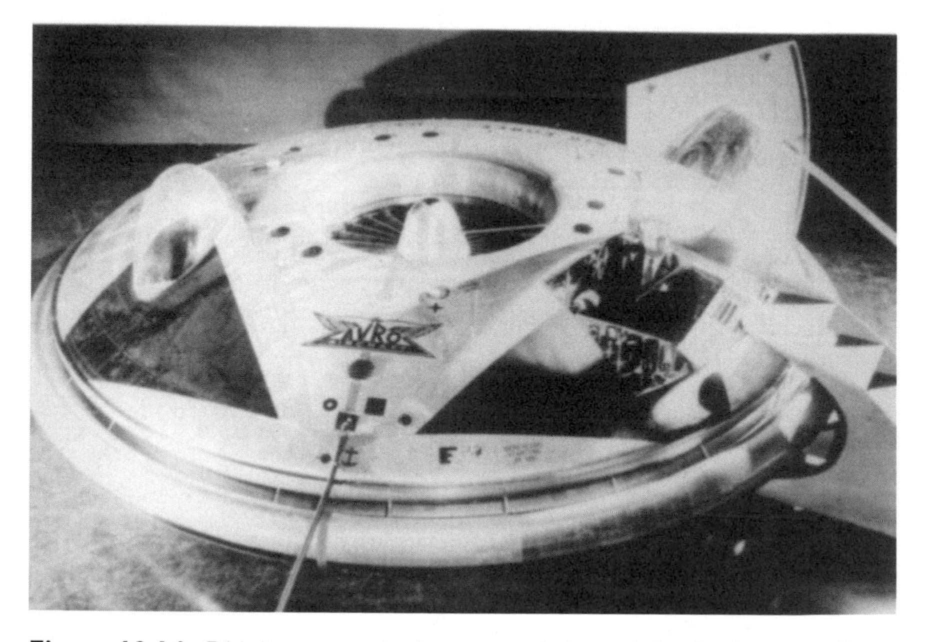

Figure 10.14. DH Avrocar: note the two cockpits and the fan in center. From photograph collection of the Royal Aeronautical Society, used with permission.

Turboplan

Switching gears now to small radio-control models, a notable spinning vehicle is the Turboplan, a radio-control toy initially designed by Heinz Jordan of Klagenfurt, Austria, and sold in the early 1980s. Two versions were sold, one of 96 cm diameter, the other a more maneuvrable 80 cm: flying weight was between 2.8 and 3.1 kg, using a 1.4 kW engine (radio-control model aircraft engines usually have the power described simply by engine capacity—in the U.S. this would be a ".61" engine, referring to capacity in cubic inches, in this case $10\,cm^3$).

In some sense the vehicle is simply a rotorcraft with a ducted fan. However, the duct itself rotates, and in many cases has enough horizontal extent to act as a wing in forward flight.

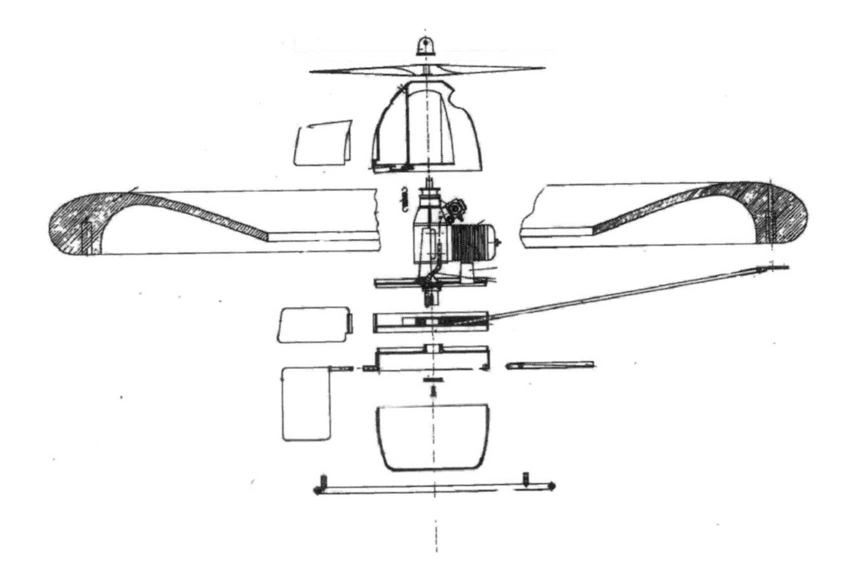

Figure 10.15. Exploded cross-section of the Turboplan. Note the shape of the duct, which guides airflow into the fan region. The concave lower surface may help confine high pressure air to provide some duct lift. Movable flaps in the lower section permit directional control.

This rotating ring-wing gives the vehicle substantial angular momentum. This momentum bias gives the system considerable stiffness (i.e., gyroscopic stability). As a result, the Turboplan is quite easy to control in hover (although less so in forward flight), which is not the case for small radio-control helicopters.

The ring is spun by means of vanes immersed in the wash from the fan—they (12 are used in one design, 8 in another) are typically mounted at an angle of about 15 degrees to vertical, and typically spin the ring to over 100 rpm (i.e., 2 revolutions per second, although the actual value depends on the details of the vanes and on the thrust being applied).

In some recent variants, the ring is simply a hula-hoop to provide gyroscopic stability (see Figure 10.17) but the thrust performance of the unaugmented fan is poor and, like the Avrocar, the vehicle is helpless out of ground effect.

Compared to a helicopter with a rotor diameter equal to the diameter of the duct, the Turboplan is a rather poor performer (e.g., in terms of the weight that can be lifted for a certain flight power). However, its performance is much better than the diameter of its small *fan* would suggest (by causing the airflow through the fan to affect a wider volume of air, the duct in effect acts as if it were an extension to the fan blades), and it is easier to control in hover than a helicopter, and reportedly can be flown in winds of 10 m/s.

Figure 10.16. Turboplan in flight. Photo by Hank Renz, used with permission.

Figure 10.17. Kelly McComb's variant of the Turboplan. Although poorer in performance than the tighter, shaped duct, this planar variant allows a good view of the control fins and vanes. Photo by Kelly McComb, used with permission.

The vertical position of the fan relative to the duct is crucial. If the fan is too far forward (high), the inward flow across the top of the wing is not set up and the wing does not contribute to lift.

The Austrian patent application features boundary layer fences to guide the flow across the surface of the ring-wing in a vortical pattern. This may influence the spin of the wing by exerting a torque on it, and by setting up a spiral airflow may affect the response time of the vehicle.

Control is effected by a pitch flap in the propwash from the fan, and a rudder. Because of the vortical flow, the response in one direction is faster than in the other.

Taiyo Edge UFX

Toys can often have exotic configurations. One example (many others exist) is the Taiyo UFX. This 200 g toy is in essence just a 40 cm six-bladed rotor, driven by two small propellers—a helicopter without the helicopter body. A small battery in the conical hub powers two motors. Even with the light body and large rotor, the vehicle can only fly for a few minutes before it needs recharging. There is no horizontal control—the user just commands the engines on and off to control the height, although if the vehicle is launched not-quite vertically, it precesses like a gyroscope and some limited directional control might be possible by modulating the thrust.

Figure 10.18. The Taiyo Edge on the author's lawn. Small propellers are driven on two of the six rotor blades.

Raytheon Whirl

Superficially similar, but vastly more sophisticated, is the "Whirl" developed by Raytheon's "Bike Shop" (a rapid development team, comparable in concept with Lockheed's "Skunk Works"). This rotary UAV is intended for long-duration radar surveillance. The idea is that a rotorcraft can remain on station without having to fly in long lazy circles. However, a UAV can have all its systems mounted on the rotor, and dispense with the inefficiency of a stationary fuselage. Although no outdoor tests have yet been reported, the intent is that the vehicle, described as a "cross between a ceiling fan and a sailboat" could hover above 10 km altitude (outside the range of antiaircraft missiles) for some four days.

The Whirl has four 3 m long rotor blades attached to a central Frisbee-like hub. The rotor is kept spinning by propellers mounted above each rotor. Its horizontal motion is controlled by rudders at the tips of each rotor, and it is control of these rudders that is the technical challenge, allowing the Whirl to tack upwind.

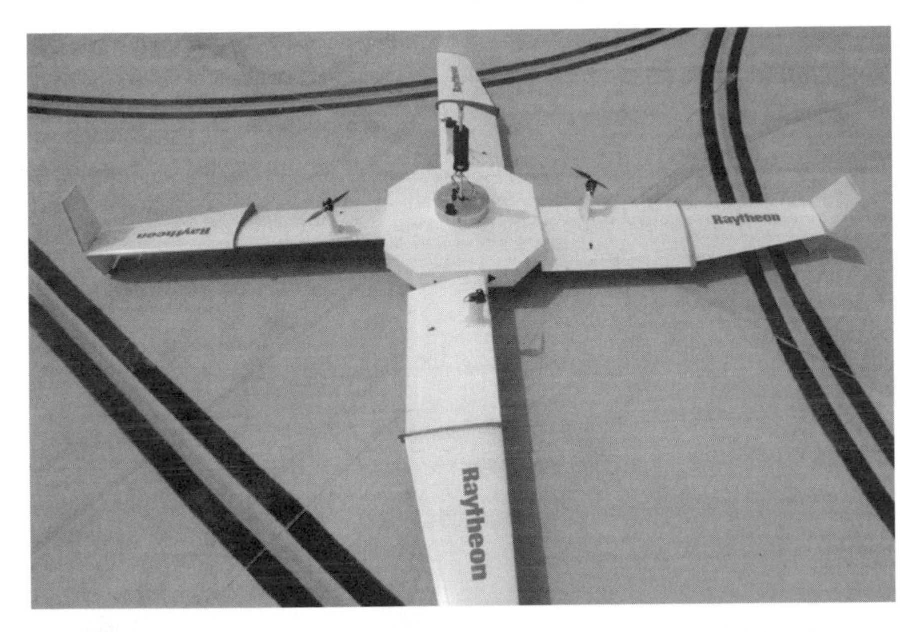

Figure 10.19. The Whirl in plan view. Photo courtesy Jim Small/Raytheon.

Figure 10.20. Photo of the Whirl at Raytheon's Tucson plant. Photo courtesy Jim Small/Raytheon.

References

Ginter, S., *Chance Vought V-173 and XF5U-1 Flying Pancakes*, Naval Fighters Number 21, ISBN 0-942612-21-3.

Lindenbaum, B., and W. Blake, *Out of the Past: Progress? The VZ-9 "Avrocar"*, Vertiflite, Spring 1998. 40–43.

McCombs, K., personal correspondence. 2004/2005.

NewScientist.com news service, Spinning spy plane could hover for days, 17:36 09 September 04.

Zimmerman, C. H., Characteristics of Clark Y airfoils of small aspect ratios, *NACA Report 431*, 1933.

Zuk, B., *Avrocar—Canada's Flying Saucer*, Boston Mills Press, 2001.

http://www.boeing.com/defense-space/ic/awacs/flash.html
http://www.fas.org/man/dod-101/sys/ac/row/an-71.htm
www.turboplan.de

11
The Boomerang

Of all the objects described in the book, the boomerang perhaps displays the most disproportionately complex behavior given its apparently simple configuration.

Boomerangs are conventionally divided—by aerodynamicists, at least—into two classes: returning and straight-flying. The former class are largely recreational, while straight-flying boomerangs (the word is derived from "bumarin," from an aboriginal tribe in New South Wales) were early hunting weapons, perhaps occasionally used in warfare.

These straight-flying boomerangs, sometimes called "kylies" or killing-sticks, are every bit as sophisticated as their returning cousins; they develop appreciable lift in flight, without the moments that lead to a curved trajectory which would be undesirable in a hunting weapon.

Figure 11.1. A conventional boomerang throw. As with strong Frisbee throws there is substantial torso and shoulder movement. The boomerang acts as an extension to the arm, and thus makes about half a rotation as it is swung through about two arm lengths. Photo courtesy of Michael Girvin.

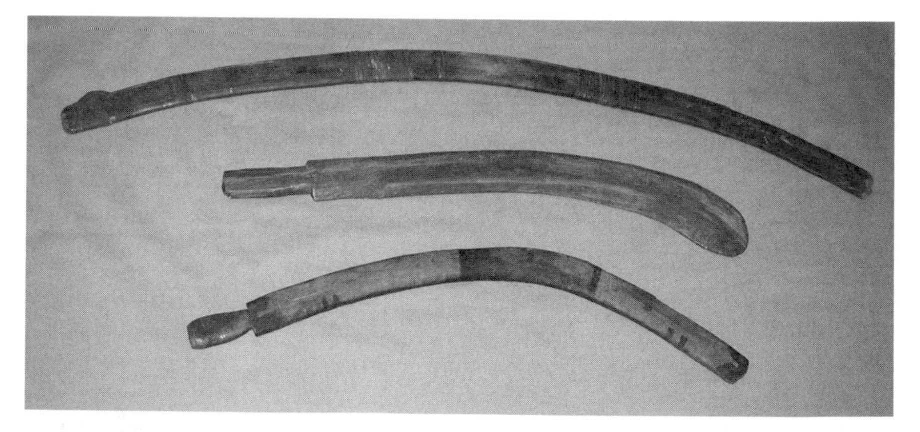

Figure 11.2. Native American rabbit sticks. Photo by Ted Bailey, used with permission.

(Musgrove (1976) suggests that the term "non-returning," often applied to these objects, is inappropriate in that it suggests that they are somehow inferior, hence his suggestion of "straight-flying.") Indeed, killing sticks can be accurate at ranges of up to 200 m, much farther than a spear can be thrown (Bahn, 1987). In this sense they represent a rather advanced throwing weapon.

Boomerang History

Boomerangs feature early in many civilizations. In addition to the Australian aborigines with whom the boomerang is now so strongly identified, examples are found among the Hopi Indians in the American Southwest, and peoples in India, Egypt, and elsewhere. North African rock paintings from 7000 B.C. show examples. In most cases, boomerangs were superseded by other weapons such as the bow and arrow as these more sophisticated technologies were developed.

Boomerangs with gold tips were discovered in Tutankhamun's tomb (Bahn, 1987)—such weighted tips can increase a boomerang's range. Wooden boomerangs have been recovered from Australian peat dated between 9000 and 10000 years before present ("BP," Luebbers, 1975) and in sand dunes in the Netherlands around 2200–2400 yrs BP (Hess, 1975). A plywood model of the latter was made and found to be of the returning type.

A remarkable boomerang, made from a mammoth tusk, was discovered in south Poland and dated to 23,000 yrs BP. (Valde-Nowak et al., 1987). This rather heavy (~ 800 g) find, with a span of some 70 cm and a chord of 6 cm at its widest point and a thickness of up to 1.5 cm, was of course too valuable to test-fly, even though it was found essentially intact. Subsequently, a replica was made from plexiglass (with a similar density of 1800–1900 kgm^{-3}) and was found to be of the straight-flying type (Bahn, 1995).

Australian boomerangs were first documented in the west in the early 19th century (the first reported use of the word may be in *The Sporting Magazine* around 1829) and became popular souvenirs. The combination of postwar lifestyle changes with expanding leisure and travel, together with the ready availability of plastics, caused them to become popular in the 1960s and 1970s. Against this backdrop, Hess (1975) provides a very extensive bibliography and review of the ethnographic literature. A good ethnographic overview, discussing regional characteristics of Australian boomerangs, is Jones (1997); a more recreational history is given by Ruhe and Darnell (1985). Ted Bailey's website www.flight-toys.com is another excellent resource.

Use of the Boomerang

Although it is found in many historical and prehistoric societies, the manufacture and use of the boomerang is best documented in Australia.

A boomerang's sharp edge can be rapidly sawed against another piece of wood to generate heat, thus (much like a fire drill) providing an ignition source. Some Australian boomerangs bear scorch marks from this process. A less esoteric but possibly widespread application of the boomerang is as a tool for digging a firepit, and as a poker. Many boomerangs discovered in the outback have tip scorching as a result.

The principal hunting method using the classic returning boomerang is indirect. The boomerang is thrown above a flock of ducks or other waterfowl. They interpret the fast-flying boomerang as a hawk or other raptor, an interpretation that may also be guided by a hole drilled in the boomerang to make a noise, or by simulated bird cries uttered by the thrower. The fowl take flight, but stay low to avoid the apparent predator. This low flight brings them in easy range of the hunter's clubs or nets. Sometimes boomerangs are used in a direct ground-to-air attack, usually at dense flocks of parrots or cockatoos.

Direct attack is of course the application of the nonreturning boomerang or kylie. Occasionally used in warfare, these are more often used to hunt small game. A particular hunting application found in a few areas of Australia is for catching fish. These boomerangs (which do not fly far, nor do they return!) are rather short in length, but heavy, so as to penetrate the water. Most likely they simply stun the fish rather than kill it outright, and are apparently used on fish at depths of about 20 cm.

The use of a boomerang as a recreational device is of course obvious. In addition to use by children, competitive displays of prowess with the boomerang (e.g., throwing to accurately return) were used in inter-tribal relations. Boomerangs are often used in dance activity, either as a gesturing implement, or in pairs clapped together as a percussion instrument. Hess (1975) also documents the use of the boomerang in fertility rituals.

In modern times, the use of the boomerang worldwide is dominated by sporting and recreational applications. Industrial production techniques and modern plastics make boomerangs with good flying characteristics very inexpensive; their manufacture is no longer only the product of intensive craftsmanship and specialist knowledge. Modern paints also make it possible to apply dramatic colors and patterns robustly enough to endure wear in flight.

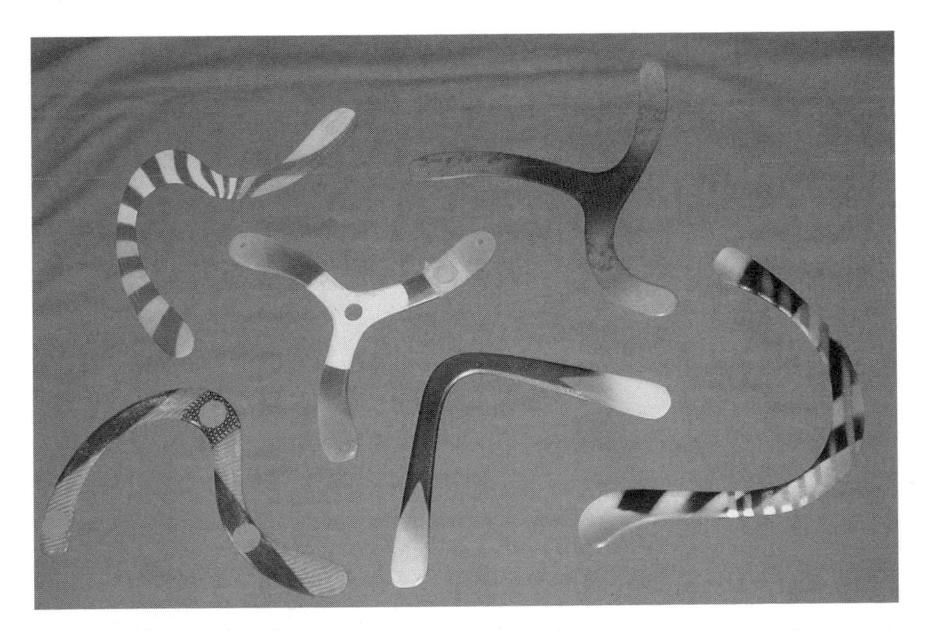

Figure 11.3. A selection of exotically shaped and colorfully decorated competition boomerangs. Photo by Ted Bailey, used with permission.

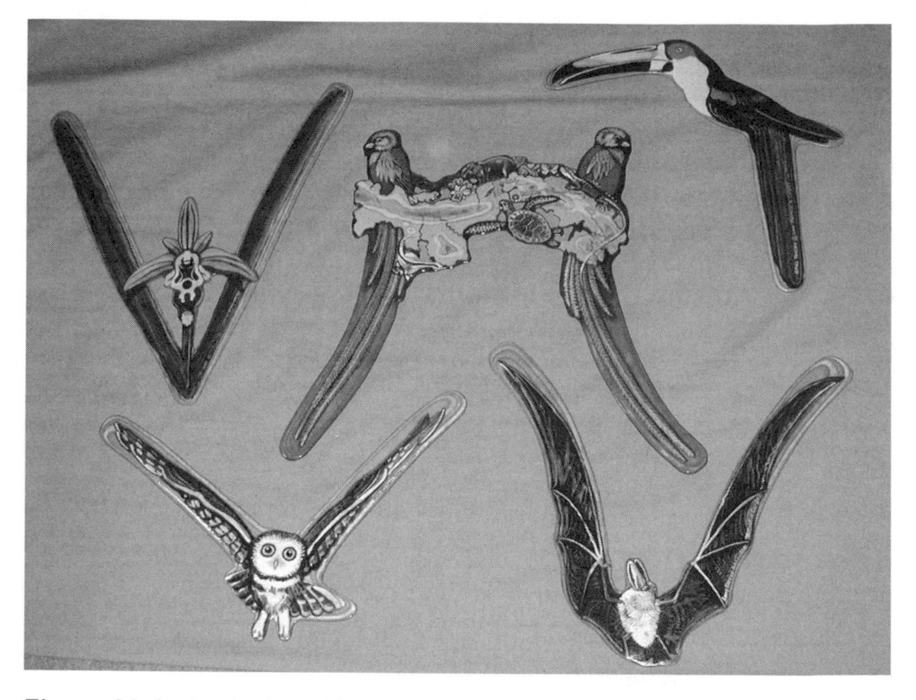

Figure 11.4. A selection of boomerangs by Cardiff Atr, shaped and decorated to resemble a flower, a bat, and one or a pair of birds. Photo by Ted Bailey, used with permission.

Boomerang Dynamics Studies

As might be expected with a design that is prehistoric, boomerang characteristics are familiar empirically, but the scientific literature on them is rather sparse. They became familiar in the English-speaking world through the explorations of James Cook, who documented their use in Australia.

Some early attempts to describe their remarkable dynamics were made, even when aerodynamics was in its infancy and flat-plate models were assumed for the lift characteristics. One significant early paper is that of Walker (1897), which cites some earlier German work, and itself offers analytical descriptions of the force distribution on the boomerang, as well as drawings of thrown trajectories. Another early work is that of Sharpe (1905), which discusses qualitatively the dynamic behavior.

This paper, while relatively short, lacks illustrations and is therefore rather heavy going.

A key feature to note about boomerangs is the basic shape. In essence the boomerang acts as a propeller—the two (or more) arms act as rotor blades as they spin to force air through the disc described by their rotation. For this to happen there must either be a twist in the boomerang (this twist attracted much early thinking before the principles of propellers were eventually understood) or it must be shaped such that the same side of both arms develops lift. This is the more usual configuration.

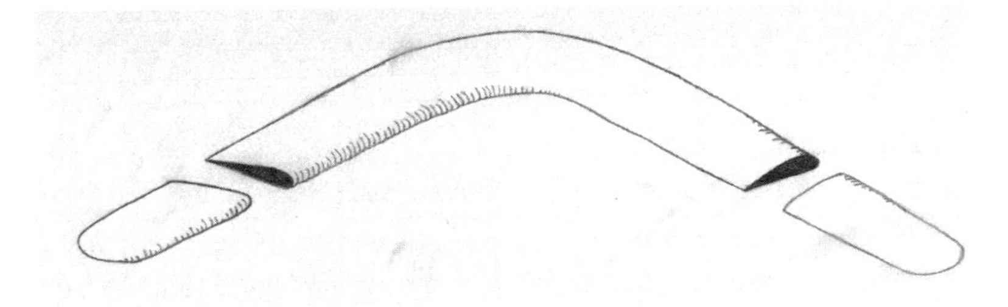

Figure 11.5. Basic shape of a typical boomerang. Usually the included angle is between 90 and 150 degrees. The section is cambered or twisted such that both arms develop lift in the same sense (in this case, an anticlockwise rotation leads to upwards lift).

These early studies captured the essence of boomerang flight, namely its operation as a propeller, combined with gyroscopic precession to yield circular flight. However, it was not until the detailed study by the Dutch mathematician Felix Hess, and his 555-page Ph.D. thesis *Boomerangs: Aerodynamics and Motion*, that boomerang dynamics were reasonably described. Hess's work presents accurate measurements of boomerang flight, and matches them with a computer model of boomerang motion. Some early parts of this work are published in abbreviated—and much more easily obtained—form in an article in *Scientific American* (Hess, 1968). The numerical model is compared with flight data, obtained by taking long-exposure photographs at night of a boomerang equipped with batteries and a light bulb.

Hess (1968) presents graphics showing how the combination of forward motion and spin gives a vertically asymmetric lift distribution (i.e., the roll moment) and how this is modified to produce a horizontal asymmetry (i.e., the pitch moment) by having an eccentric wing.

This force distribution is illustrated in Figure 11.7, which shows the distribution of aerodynamic forces, on a spinning wing at various advance ratios, for a radial wing and an eccentric wing. Naturally the forces are greatest at the wingtip that is moving into the flow; at parts of the rotation where the circular motion cancels out the forward motion, the forces are zero. Although there will be some lift produced where the flow is backwards (depending on the twist or camber of the aerofoil), this will typically be rather low compared with the advancing blade. The advancing side clearly (top figure) gives an asymmetry about the direction of motion, leading to a roll moment which precesses the boomerang around to make its circular path. This asymmetry decreases with advance ratio—a sufficiently fast-spinning wing will essentially feel no effect of forward motion.

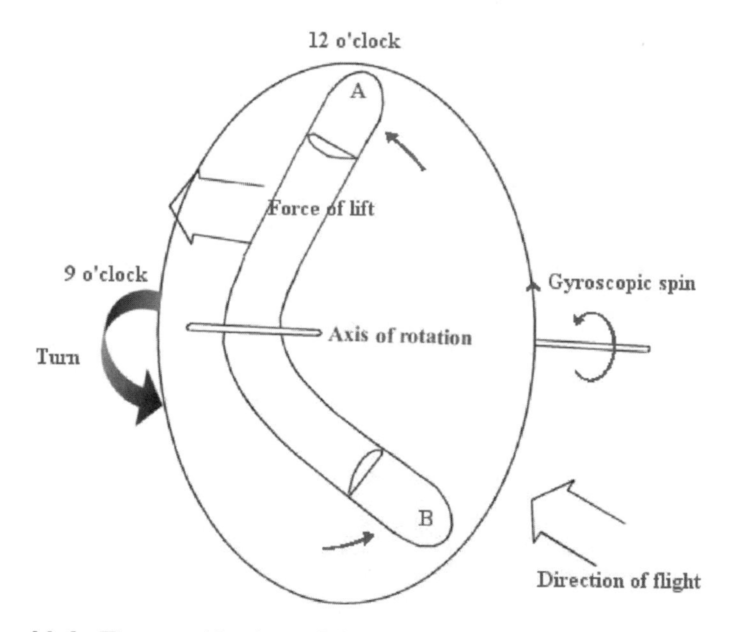

Figure 11.6. The combination of forward flight plus the forward rotational motion of the upper arm gives more lift on the upper arm than the lower. The net lift therefore acts above the axis, leading to a rolling torque. This rolling torque causes the boomerang to precess its spin axis in the horizontal plane.

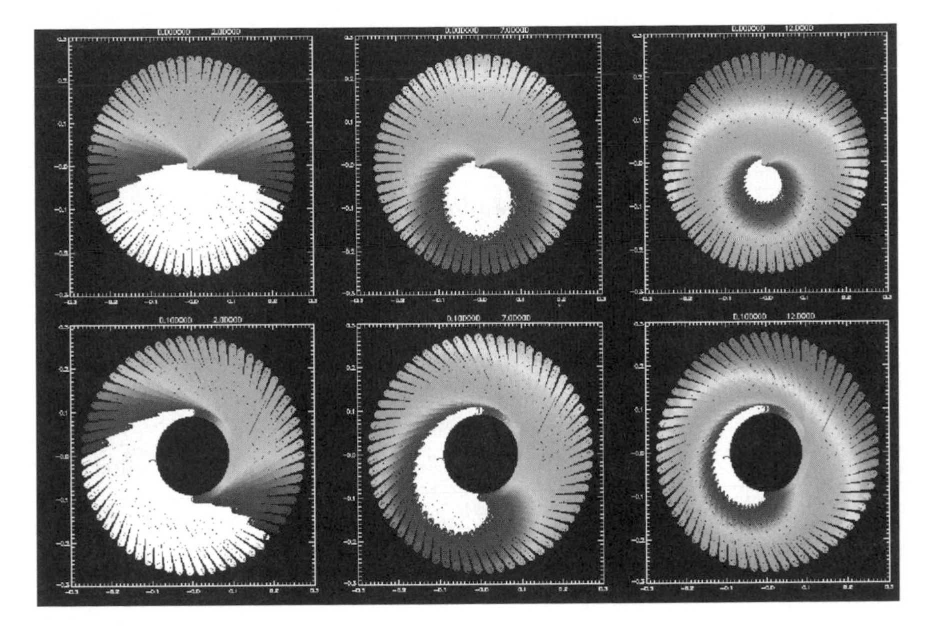

Figure 11.7. The dynamic pressure distribution on rotating wings. Upper panels are for a purely radial wing (moving to the left, rotating anticlockwise). As expected, the dynamic pressure is stronger on the upper half than the lower half, leading to the roll torque which precesses the boomerang flight into a circle. The lower panel shows an offset (i.e., not perfectly radial) wing, as is typical for a boomerang: in addition to the upper/lower pressure asymmetry, there is a left–right asymmetry—this produces a pitch torque which causes the boomerang to "lie down" into a horizontal plane.

If it is assumed that the component of relative air velocity along the length of the wing does not contribute to lift (i.e., the spanwise flow), then it can be seen that an eccentric wing develops an asymmetry about the orthogonal axis, and thus a pitch moment.

Interestingly, Hess also presents an example of a different geometry, namely a boomerang with 8 radial arms (and thus no eccentricity) which nonetheless still "lies down." In this instance, the blades are close enough together that the pitch moment is produced by wake effects.

Important though the 1968 paper is, it does not do justice to the breadth of work in the thesis, which includes much more detailed mathematical treatment of the aerodynamics and dynamics, as well as presenting wind tunnel and water tunnel measurements of boomerang force and moment coefficients. Additionally, there is a very extensive classified survey of the previous literature. Even the streak photo experiments are more advanced than the 1968 ones, with a "time pill" flasher circuit wired to the bulb to provide time information on the photographs. The photographs are taken in pairs, to provide a 3-dimensional perspective on the trajectories, and a stereo viewer is even provided in the back cover of the thesis!

Musgrove (1974) describes a mechanical boomerang-thrower, developed as a series of undergraduate research projects to launch boomerangs at known speeds, angles, and rotation rates. To the present author's knowledge, no results from this activity have been published.

King (1975) presents a simple analytic analysis of boomerang turning behavior, together with descriptions and flight data from a series of undergraduate flight experiments. One notable innovation developed by his students is a whistle incorporated into a boomerang, enabling it to be tracked acoustically, using a bowl as a parabolic reflector for a microphone.

Thomas (1983) gives a comprehensive and readable overview of the boomerang, including its dynamics as well as some details on the modern sport. Newman's 1985 paper is a short but particularly clear analysis of boomerang dynamics and aerodynamics, including the final "hovering" phase. He offers various dimensional arguments to show that small boomerangs are harder to design than large ones, and that they should be dense and have a high aspect ratio.

All of these works focus on a lumped-parameter approach, namely that the flight behavior of the boomerang can be adequately described by properties averaged over a whole rotation. A number of computer simulations of boomerang flight have been developed (one is wxBumms by Georg Hennig, an adapted version of which was used to make some of the figures in this chapter) using Hess's approach. Measurements (see later this section) show that forces and moments vary dramatically with rotation phase, and so care must be taken in using average properties. The most recent theoretical work begins to address the rotationally resolved aspects of the dynamics.

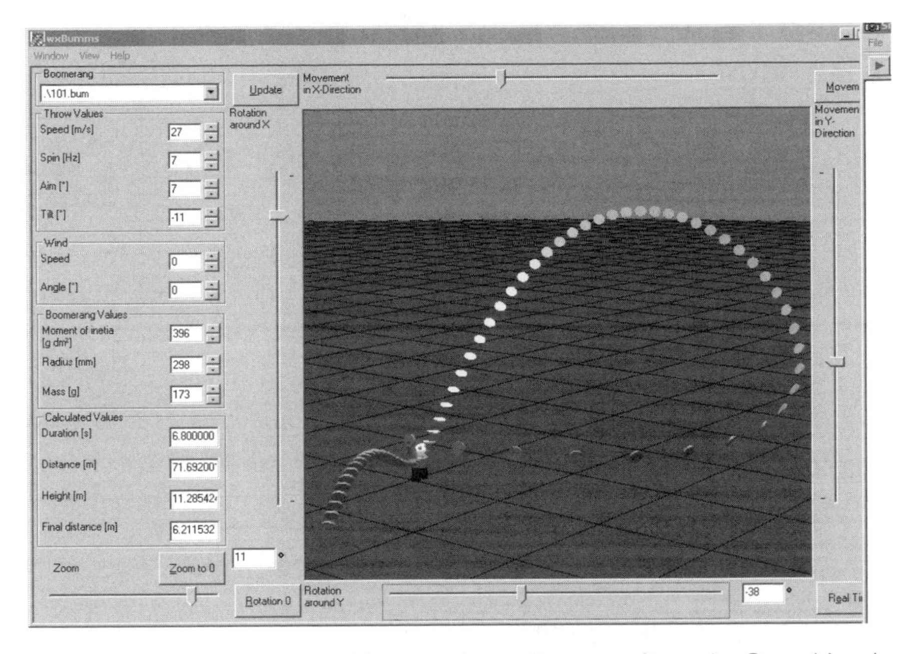

Figure 11.8. A screenshot of the very nice wxBumms software by Georg Hennig. The throw parameters can be adjusted by mouse clicks and the changes to the trajectory immediately seen from a variety of perspective angles. The boomerang's spin orientation is shown as a disc every tenth of a second of flight. Spacing between the snapshots gives an indication of the flight speed. The grid pattern on the ground is of 5 m squares. This experiment shows that one can select the parameters just right for decapitating the thrower about 5 seconds after launch!

More recently Battipede (1990) has described certain aspects of boomerang flight, with reference to the rotorcraft literature and inflow effects. While this may well be an important aspect of the flow around a boomerang, this numerical study (a graduate student project) is not compared with any real flight data. It is notable, however, in pointing out the potential importance of the cross terms in the inertia tensor for real boomerang shapes.

A considerable advance—perhaps the most significant development since Hess's work—is offered in two recent papers (Azuma et al., 2004; Beppu et al., 2004). These perform simulation of boomerang flight, but by explicitly integrating the aerodynamic forces along the span of the boomerang for timesteps much smaller than the rotation period. These studies show the instantaneous variation in angles, and indeed in the spin rate itself, and relate the overall flight performance directly to the aerodynamic properties of the wing section chosen.

Until this recent work (which was published as this book was being completed) the step of making the connection between wing section and overall flight had not been attempted, except via the indirect step of developing spin-averaged properties. It may be that the practical difficulties of making wind tunnel measurements on a rotating body which has no substance at its center of mass (unlike the Frisbee, to which a rotating sting can be attached at its center) have been a significant impediment to progress in this direction. Hess (1975) does make some measurements, with a special fixture to attach the boomerang to a rotating sting, but it is not clear how much this fixture may affect the measurements.

Boomerang Architecture

The canonical boomerang is angular or crescent-shaped in planform. In most recreational boomerangs, the two wings are of approximately equal length. For the boomerang to be effective, the two wings must both develop lift, as if the boomerang were a propeller, when it spins in one direction, the direction determining the "handedness" of the boomerang. "Right-handed" boomerangs are thrown with the "upper" surface of the boomerang pointing left—the upper surface points towards the thrower's head, and to the center of the circular flight path.

For both wings to develop lift, their aerodynamic surfaces must be shaped accordingly. In the (rather bad) case where the wings are perfectly flat or at least uncambered, the boomerang must have twist, such that both wings encounter a positive angle of attack and thus generate lift in the same direction. It was thought in the early days of boomerang study that such twist was essential.

In fact it is not. If a thicker aerofoil is used with, e.g., a flat base and a curved upper surface, then lift is positive at zero incidence, and this will apply to both wings. A similar result pertains if the airfoil is cambered in the same direction on both wings.

The classic type of boomerang is simply angled, or perhaps has a slightly reflexive "Omega" planform—the shape displacing the center of pressure from the center of mass to yield the desired moments during flight (see below). Some other designs are more radially symmetric, forming a three or more pointed "star" shape (the four-armed cross designs come under this category). Several other permutations are possible, many resembling letters of the alphabet (N, T, W, etc.). A final variant is the Aerobie Orbiter, which has an open triangular planform, allowing it to be caught by placing a hand (or foot) in the "hole." In this example and many others, there is a twist applied to the tip to manipulate the lift distribution along the span.

Figure 11.9. A boomerang being caught by the author at the end of its flight (Arizona's Sentinel Peak is in the background). Notice that at this late stage in the flight, the boomerang is in a near-horizontal orientation, in contrast to the near-vertical throw orientation (Figure 11.1). Image courtesy of Curtis Cooper.

Traditionally boomerangs are carved from a carefully chosen piece of wood; more recently plywood has been a favored material, especially for homemade designs. The relevant material properties beyond manufacturing considerations are the material's density and its robustness to damage in flight. A badly thrown boomerang can hit the ground distressingly fast. For indoor use, lightweight boomerangs are sometimes made from a low-density foam material or balsa wood, both to minimize mass and moment of inertia to permit a small turn radius, and to minimize potential damage to items or people indoors. A quick and dirty boomerang can be made by strapping two wooden or plastic rulers together with rubber bands or tape (some varieties of ruler have a beveled edge which defines an aerofoil shape).

Figure 11.10. A four-bladed paper boomerang held by space shuttle astronaut John Casper on STS-54 on January 13, 1993. He threw this 2 g, 18 cm boomerang in the mid-deck of the shuttle—unsurprisingly, it hit the wall in this cramped area.

Beppu et al. (2004) made a series of flight trials of a set of elbow-shaped boomerangs with different joint angles. As the joint angle is increased, the boomerang tends to have a higher moment of inertia about its spin axis, which tends to improve its flying characteristics. However, as an angle of 180 degrees is approached, the transverse moment of inertia of what is now, in essence, a rectangular plate like a ruler becomes very low, and so it is all too easy for it to "flip" and begin tumbling. Thus for more usual angles of around 90 degrees when the mass distribution more closely resembles that of Frisbee, the moments of inertia are such as to favor stable flight. Beppu et al. (2004) found that only boomerangs with joint angles of between 40 and 120 degrees could be flown stably. It may be that downwash effects become more significant for small included angles—the 30-degree elbow angle boomerang was unflyable.

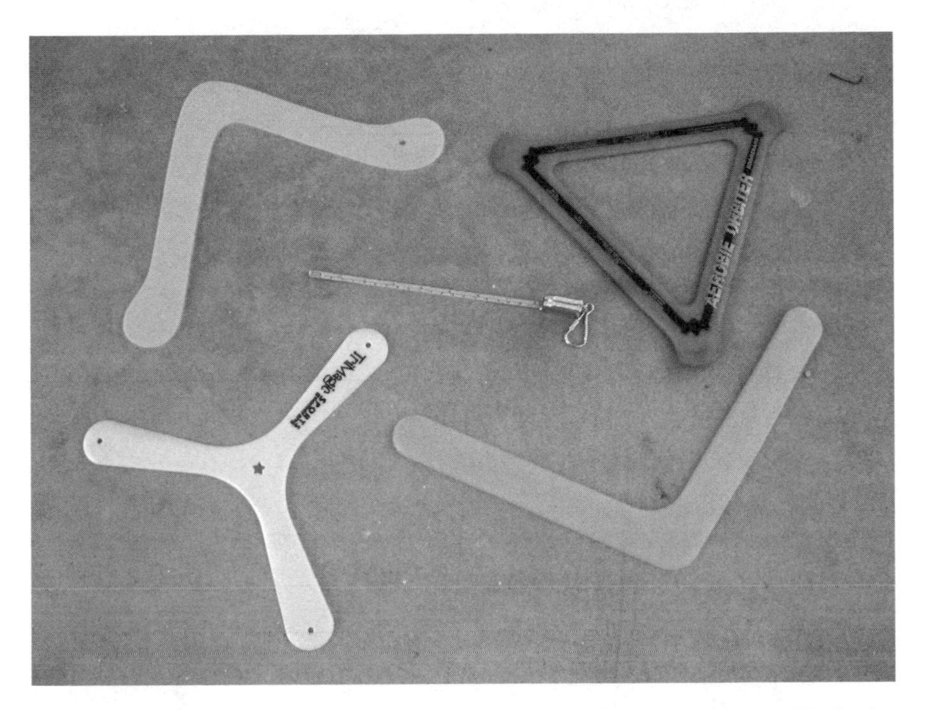

Figure 11.11. The author's collection of cheap plastic boomerangs. Clockwise from upper left are the polypropylene Pro-Fly by Eric Darnell, the Aerobie Orbiter designed by Alan Adler, the Spin-Bak traditional, and a Trimagic.

Some boomerangs have one arm substantially longer than the other. This is particularly the case with MTA (Maximum Time Aloft) boomerangs, and is also characteristic of many Australian aboriginal examples. The long wing gives a large effective moment arm (much like a slingshot or spear-thrower) to permit a higher launch velocity. Also, providing the stability concerns mentioned earlier can be addressed, the moment of inertia of an asymmetric straight-ish boomerang will be maximized for a given mass; the rotational kinetic energy is what maintains the hover, which is the most important phase for maximizing flight time. A further point is that a long wing gives a higher aspect ratio, which Newman (1985) points out is a key parameter for maximizing the number of turns made by the rotor before motion stops.

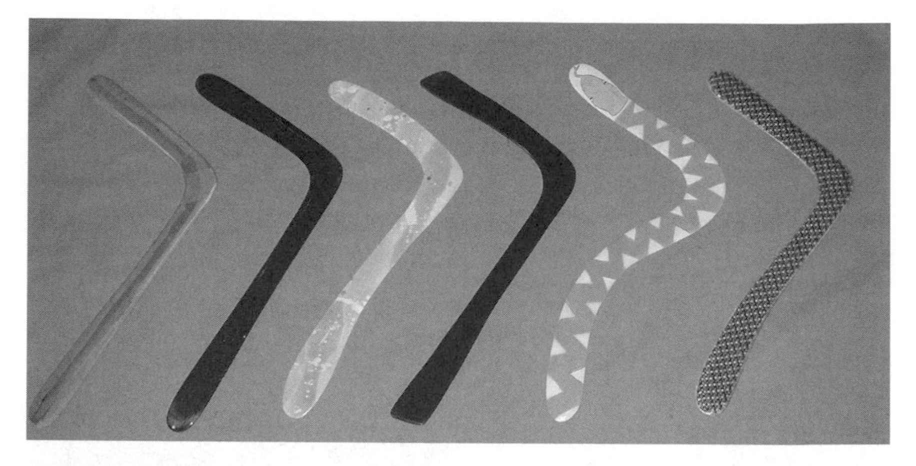

Figure 11.12. A selection of modern Maximum Time Aloft (MTA) boomerangs. All have the characteristic asymmetry in arm length. Photo by Ted Bailey, used with permission.

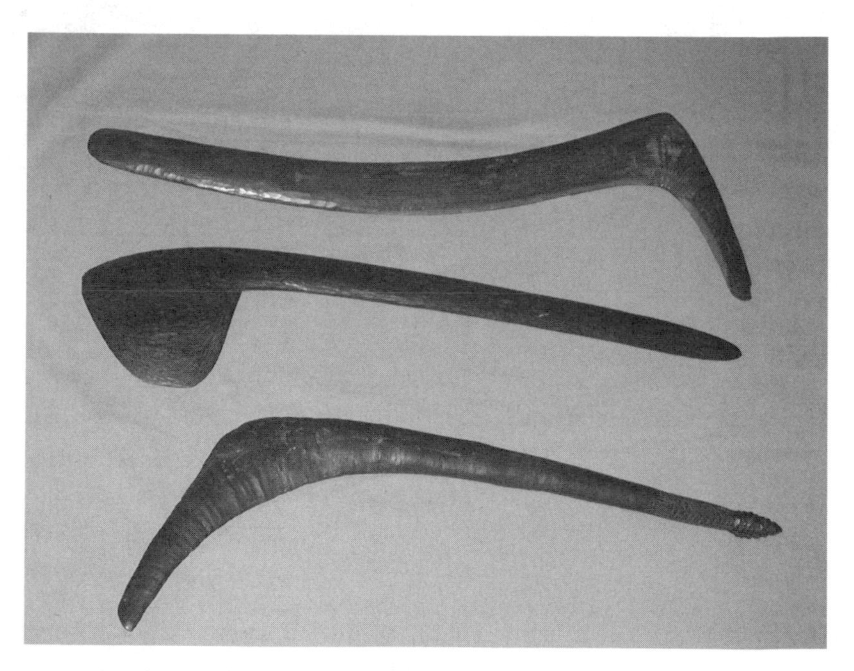

Figure 11.13. Unusual Australian aboriginal throwsticks. Photo by Ted Bailey, used with permission.

Mass distribution is a critical factor for boomerangs, to adjust the center of mass and to change both the mass and the moment of inertia. Large moments of inertia are generally favorable for longer flights (see below), and for this reason masses are often added to the wing tips.

A final architectural aspect is material and coating. Boomerangs are characteristic of dry areas, perhaps in part because the flight characteristics are very sensitive to wing twist, which may be induced in wood by moderate or high humidity. Recreational boomerangs tend to be lacquered or otherwise protected from humidity, or made outright from plastic materials. The plastic should permit some bending and retain it (to permit "tuning") without deforming too much in regular use.

Returning Boomerang Flight: Essentials

In a classic boomerang flight, the article is thrown with its principal plane inclined outwards by about 20 degrees from vertical. The projection onto the horizontal plane of its flight is approximately circular, with a diameter of typically 30 m; its flight path is initially inclined such that it climbs perhaps 10 or 20 meters into the air. At the end of its circular arc, its forward motion has decayed and the boomerang falls to the ground.

The rising and curved path follows simply from the development of lift: most of the lift is projected onto the horizontal plane, causing the article to veer inwards (i.e., to the left, for a right-handed throw). With the initially high forward velocity, the vertical component of lift exceeds the weight of the boomerang, and causes it to accelerate upwards.

The clever part of boomerang design derives from the aerodynamic moments. As a thought experiment, consider a throw unaffected by gravity, and with no turning moments. The spin plane would remain fixed in inertial space and thus the boomerang would accelerate to the left until its leftward motion introduced a sufficiently negative angle of attack that the lift fell to zero.

A circular flight path requires that the spin axis be precessed anticlockwise, as seen from above. This is accomplished by a roll moment which is due to the upper wing experiencing a higher airspeed due to the spin—its circumferential velocity adds to the forward velocity of the boomerang, while the lower wing's circumferential speed subtracts from it. This causes an inwards roll; this incremental angular momentum vector points backwards along the direction of flight. Adding this to the spin angular momentum causes the rotation of the latter in the horizontal plane.

Thus far the story is straightforward, and any rotor will experience the same sort of effect—a purely linear wing, for example. A common model for this behavior is a cross-shaped boomerang. This is easily fabricated from balsa wood, or even two plastic rulers bound together with adhesive tape or rubber bands.

A neat analytic result follows from this paradigm of boomerang flight, namely that the radius of the circular flight depends only on three fixed parameters: the lift coefficient, moment of inertia, and span (Hunt, 1999):

$$R_F = \frac{4I}{\rho C_L \pi a^4},$$

where R_F is the radius of the circular flight, a is the span of the boomerang, C_L the lift coefficient, and I the moment of inertia about the spin axis. This result relies on C_L being invariant, whereas in reality it will depend upon the angle of attack of the throw (although this is not an easy parameter for a thrower to adjust—generally boomerangs are launched at zero angle of attack).

A neat feature of the classical boomerang is its tendency to "lie down"—the spin plane is initially inclined 20 degrees from vertical, but over the course of the flight it rotates outwards such that the boomerang is more or less horizontal at the end of the flight. This rotation of the angular momentum vector from near-horizontal to vertical requires a pitch-up moment, in essence due to the lift acting forward of the center of mass.

Usually this behavior requires the characteristic angled shape of a boomerang. With such a shape, the apex is forward of the center of mass, and the wings are no longer radial to it, but are eccentric. A simple calculation of the square of net velocity (forward plus spin-induced) shows how the center of pressure moves forward as a result.

Note, however, that not all boomerangs can be explained this way. Hess (1968) shows a radially symmetric 8-bladed boomerang (in essence, a throwing star) which also "lies down". In this case, the forward displacement of the lift must be due to the trailing side of the boomerang operating in the wake or downwash of the leading side. As with the Chakram or Aerobie flying ring, this downwash reduces the effective angle of attack of the trailing side, and thus the lift force from it. The reduced lift on the trailing side therefore displaces the center of pressure forwards and thus precesses the spin axis towards the vertical.

A very crude calculation shows the relative roll and pitch moments as being roughly in the ratio of 2 to 1 (this was shown by Walker in 1897): the spin plane is precessed horizontally through something over 180 degrees, while its precession to near vertical involves around 90 degrees. The reality is of course rather more complicated, since the spin rate decreases through the flight, and the horizontal precession may be as much as 360 degrees or more.

Note that the horizontal precession (i.e., the roll moment) is a function of the spin rate and flight speed which determine the airspeeds over the upper and lower wing, while the pitch moment that precesses the angular momentum vector vertical is largely a geometric property.

Infinite Boomerang Flight

Hess (1975) performs some interesting numerical experiments, one of which is to consider the flight of a boomerang thrown upwards, but then allowed to fall downwards without limit, as if the ground were not present. This situation is in fact of interest were a boomerang dropped from an aircraft, perhaps to descend through the atmosphere of another planet. (In this sense, the boomerang can be considered a special case of a samara.)

As is often observed in a conventional boomerang throw, the first (usually upwards) loop is anticlockwise (as seen from above, for a right-handed boomerang), and then the forward flight slows and the boomerang reverses its flight direction, usually to hit the ground soon thereafter. Without the ground, this reversal completes itself and the boomerang begins a clockwise spiral. Like a coil spring, this spiral maintains a constant radius and "pitch" (i.e., vertical interval between successive loops).

Hess makes the observation that the familiar anticlockwise boomerang flight, with its reversal, is essentially just a transient step towards a terminal clockwise spiral; the latter is the "natural" state of the boomerang.

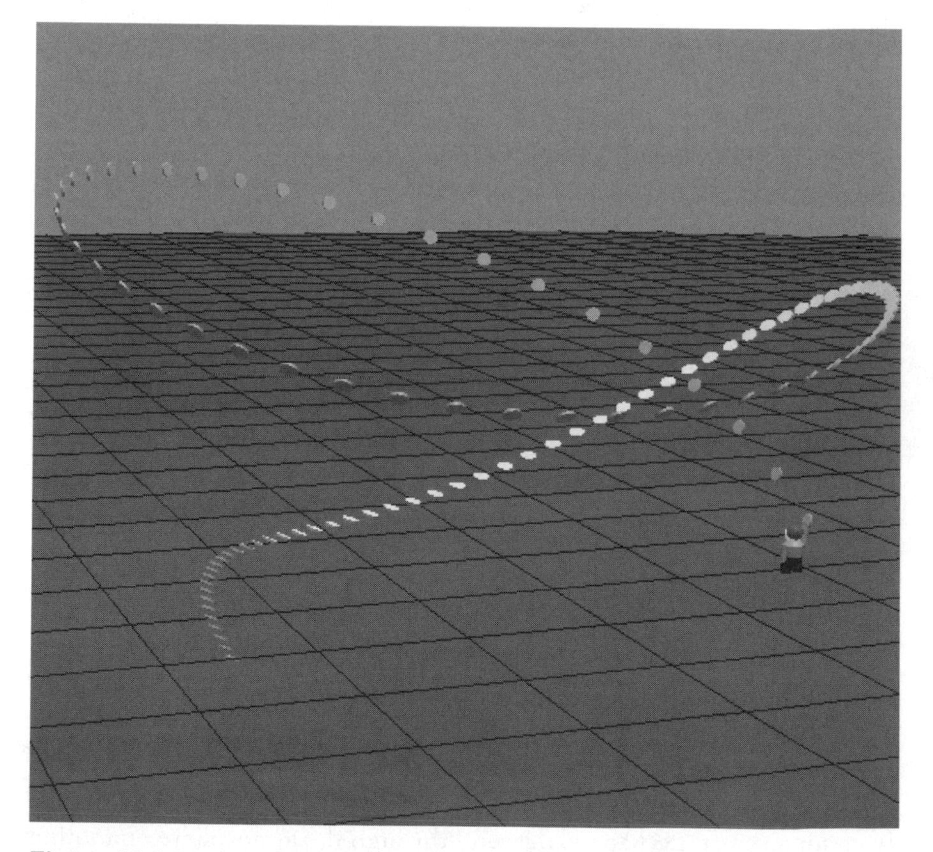

Figure 11.14. This rather hard throw shows how a second loop can begin during long flights, starting towards an infinite spiral like a samara, were the ground not to get in the way.

Trends with Throw Parameters

Increasing the flight path angle (the angle made by the velocity vector with the horizontal) causes a boomerang to reach its maximum altitude more quickly, yet surprisingly it tends to fall down more quickly too.

Increasing the angle of attack increases the lift coefficient, and thus (following the simple model above) makes the radius of the flight path smaller, making a tighter loop. One way of increasing the angle of attack fairly early in the flight is to add weights to the inboard part of the boomerang, i.e., to increase the boomerang mass, without increasing the

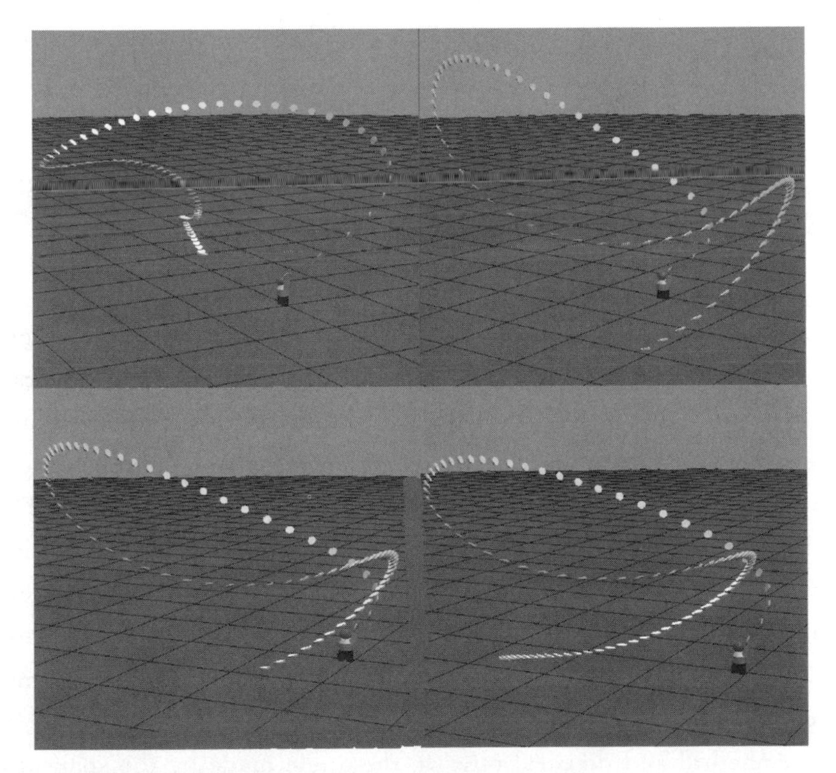

Figure 11.15. Composite of boomerang trajectories for a single boomerang with only modestly different throw parameters. Clockwise from top left, the parameters are $v = 26\,\text{ms}^{-1}$, $\omega = 9$ rps, tilt = 29°, aim = 8°, a throw that descends very slowly but ten meters away; $v = 30\,\text{ms}^{-1}$, $\omega = 9$ rps, tilt = 29°, aim = 8° swoops just in front of the thrower to land behind; $v = 30\,\text{ms}^{-1}$, $\omega = 9$ rps, tilt = 16°, aim = 8° is a more gentle swoop; $v = 26\,\text{ms}^{-1}$, $\omega = 9$ rps, tilt = 11°, aim = 8° should be easy to catch.

moment of inertia substantially. This retains the same yaw rate (i.e., the spin axis precesses anticlockwise as before), but the velocity vector is rotated anticlockwise more slowly, since the lift now has to accelerate a larger mass. The difference between the two rates yields a rapidly increasing angle of attack.

Increasing the roll angle (i.e., launching the boomerang in a more horizontal plane) causes the lift generated by the boomerang to have a stronger vertical component, accelerating the boomerang into the sky.

Increasing the spin rate has only a modest effect. This increases the pitch moment (since the advancing and receding wings of the boomerang have a larger speed difference), but at the same time increases the angular momentum. To first order, then, the effects are the same magnitude and cancel out, although for very high spin rates this will not be the case, and the pitch moment will grow faster.

Trends with Aerodynamic Properties: Tuning

The aerodynamic properties of boomerangs are of course dictated by the aerofoil shape and the planform as constructed. However, the flight characteristics can be modified significantly by small deformations (twisting and bending) done in the field—"tuning." Often recreational boomerangs are supplied with instructions on how to perform these adjustments, but the aerodynamic basis for them is rarely given. The choice of plastic material used for boomerangs is vital in determining how well this tuning works—the plastic must be soft enough to be bent by hand, yet the viscoplastic properties must be such that it doesn't bend itself back over too short a time. Traditional boomerang tuning sometimes involved heating the wood over a fire.

Anhedral and dihedral refer to the angle made by the span of a wing with the horizontal: dihedral wings point upwards from root to tip, forming a "V" shape and typically give an aircraft better stability in roll. Anhedral wings form a "Λ" and tend to have the opposite effect. Anhedral is used on fighter aircraft which need manoeuvrability rather than stability and on high-winged transport aircraft which already have substantial "pendulum" stability.

Applying dihedral (by flexing the wingtips upwards) on a boomerang tends to have the effect of inducing high, hovering flight. This requires the boomerang to "lay over" more quickly—in other words the pitch-up moment is enhanced. This is probably via the airflow hitting the underside of the wingtip at a steeper angle when the tip is pointing forwards. Conversely, applying anhedral yields a lower flight, with later lay-over.

Another field adjustment is blade twist, to increase the angle of attack of the blades throughout their revolution. This of course increases the lift coefficient and thereby leads to flight in a tighter circle.

Straight-Flying Boomerangs

As pointed out by Musgrove (1975), although a straight-flying boomerang may superficially appear simpler than the returning type, consider the fundamental feature of a spinning wing. The advancing wing will experience a higher airspeed and thus a stronger lift than the receding one, and thus a killing stick thrown in a horizontal plane will flip over. Suppressing this tendency thus requires nulling the roll moment via tuning of the lift distribution, which is accomplished by twisting the blades such that the outboard part of the span produces negative lift.

This measure may appear inefficient, but recall that the majority of the lift force developed by a returning boomerang is expended in providing centripetal acceleration to create the circular trajectory. To first order, the ratio of the horizontal component of lift to the vertical component required to balance its weight is the tangent of the boomerang's inclination to the horizontal, or for 70 degrees, around 2.7. Thus, a boomerang that does not need to make a circular flight can afford to generate three times less lift (and correspondingly three times less drag).

The blade twist (perhaps combined with a change in section along the span) yields a net positive lift—the inboard positive lift outweighs the outboard negative. However, the roll moment can be made very small, since this is the integral of the lift at each part of the span multiplied by its distance from the center of gravity, such that the outboard negative lift has greater leverage.

Figure 11.16. African throwsticks. The central one may be deliberately shaped to resemble a gun. Photo by Ted Bailey, used with permission.

Boomerang Experiments

My first experiment was simply to install flashing LEDs (see appendix) in a commercially available boomerang, the Darnell Pro-Fly procured from Ted Bailey (www.flyingtoys.com—price was about $7). This has a conventional angular/omega shape, weighing about 70 g with a span of about 28 cm. A green LED was mounted close to the apex, and would thus closely trace the course of the boomerang's center of mass. Red LEDs were mounted close to the tip of one arm, and thus would describe a much wider cycloid.

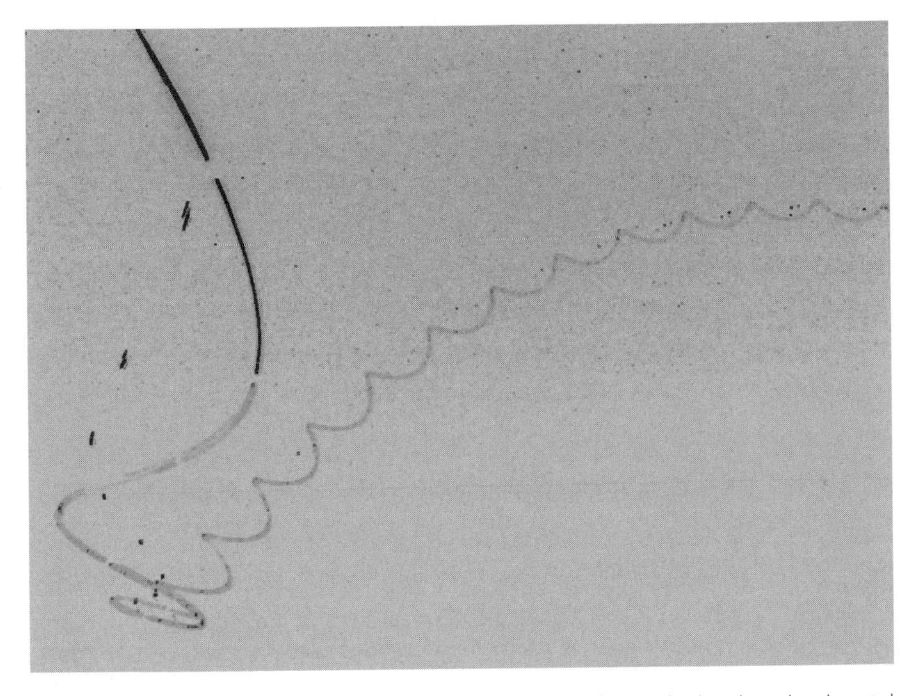

Figure 11.17. Trace of a boomerang thrown flat (i.e., spinning in a horizontal plane, like a Frisbee) over the camera. The short streaks are the LEDs near the center of the boomerang, describing a slightly snakelike pattern as the boomerang flies forward. The longer trail is of one arm tip of the boomerang—after a couple of revolutions, the boomerang turns upwards and to the right. The short breaks in the trail correspond to the ~ 10% of the time when the circuit turned off the tip LED and turned on the central LEDs. Image inverted for clarity.

The next step was to embed a microcontroller, accelerometer, batteries, etc. (see Appendix 1) in the boomerang. The 15-second sampling period afforded by the microcontroller is well suited for boomerang flights, which are usually a little under 10 seconds. The installation was a little more challenging than for the Frisbee experiments, in that the boomerang has a relatively thin section. Cavities were milled out (actually a somewhat messy process, since the boomerang is made from a thermosoftening plastic which tended to melt during milling) and components embedded in hot glue.

The accelerometer (an ADXL210; see Appendix 1. A $10g$ range was needed for boomerang flights, which experience rather more violent accelerations than do Frisbees, for which a $2g$ range is generally adequate) was mounted such that accelerations in the forward and transverse directions could be measured. The "forward" measurement (*Y*) is one essentially of drag, at least at the early part of the flight at low angle of attack and the transverse one (*X*) of lift. The accelerometers were sampled at about 30 samples per second each.

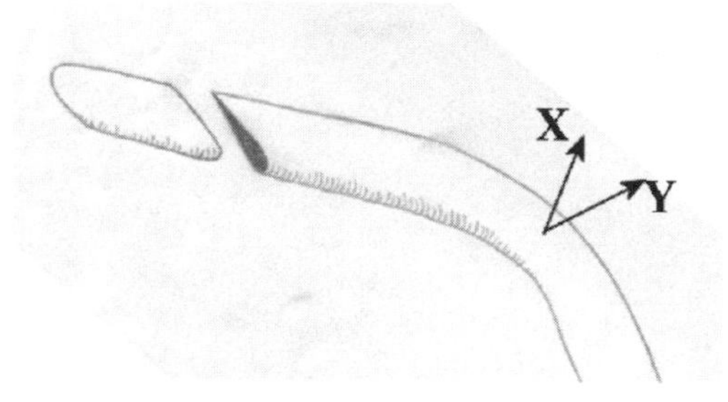

Figure 11.18. The *X* direction measured with the instrumented boomerang is parallel to the spin axis—essentially the boomerang's lift. The *Y*-axis is the nominally forward direction, although of course it is swept around to point alternately forward and backward.

Initial experiments with the microcontroller flashing some ultra-bright LEDs were not very successful—the high current needed for the LEDs tended to cause the voltage on the microcontroller to drop and thus "brown out" or reset. A separate battery was therefore installed, and the LEDs run continuously.

Rather than install marking lights in the throwing field (Hess used special lights as fiducial markers in his photographs), clear nights were chosen. Light pollution is modest in Tucson, and it was thus possible to use the constellations in the night sky as angle calibrations. Another possibility would be to use existing regularly spaced lights such as streetlights as markers.

There is weak spin modulation of the transverse accelerometer signal, superimposed on a smooth parabolic decline. A striking feature, however, is the sharp peak in lift—such, indeed, that the accelerometer was over-ranged. This peak is not associated with the launch—the lift is clearly near zero at launch. This therefore shows the increase in angle of attack in the first few fractions of a second of flight—clearly the precession torque rotates the spin vector (anticlockwise as seen from above) faster than the lift accelerates the boomerang to the left.

The forward (Y) accelerometer shows a periodic pattern, as might be expected (to first order, one revolution should be the same as the next). Unlike Frisbee flights, however, this pattern is quite asymmetric, reflecting the lack of symmetry of the boomerang itself. In the early (high speed, modest incidence) part of the flight, the pattern is double-peaked. Later, the curve has a single peak per cycle. Note that the Y accelerometer is not symmetric about zero, but in this case has a negative offset. This is due to the centripetal acceleration due to rotation, since the center of mass is well behind the accelerometer.

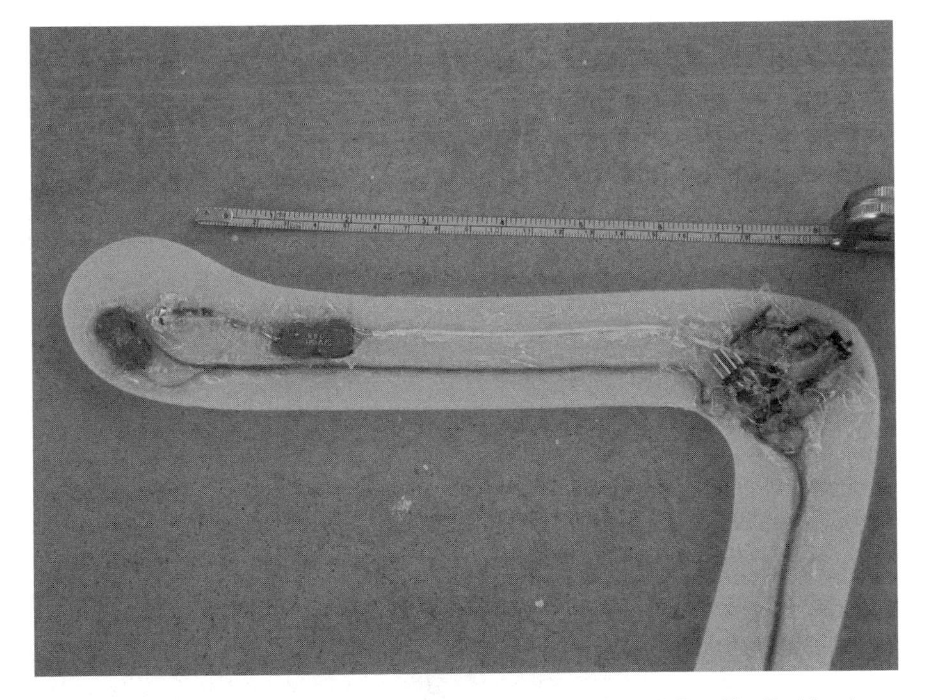

Figure 11.19. Close-up of the author's modified Darnell Pro-Fly. Cavities have been cut out in the plastic and batteries, LEDs for tracking, and a microcontroller and accelerometer have been inserted. This boomerang yielded the flight path and data in Figures 11.20 and 11.21.

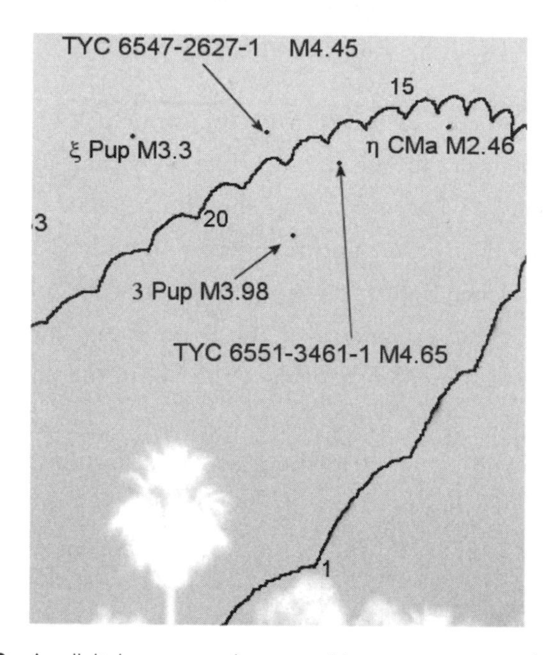

Figure 11.20. A digital camera image with contrast stretched and colors reversed—the silhouette of a palm tree appears to the left as a white shape. The cycloidal path of the LED on one arm tip has been traced over for clarity. Stars in the background have been identified by comparison with a star atlas and provide a convenient angular reference. Numbers on the flight path indicate the number of revolutions taken.

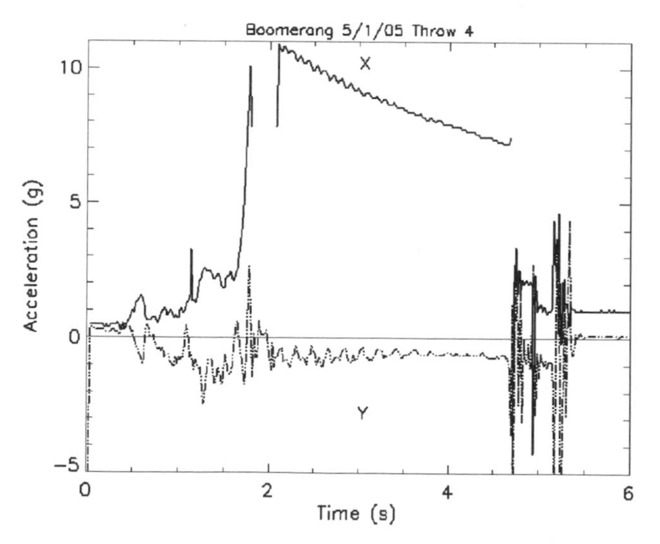

Figure 11.21. Accelerometer record from a boomerang flight. The *X* curve is the acceleration normal to the plane of spin (i.e., "lift") while the *Y* axis is in the spin plane—something like "drag." The boomerang is released at ~ 0.1 s and makes a 5 s flight. Notice that the *Y*-axis record has an irregular spin modulation, indicating that the forces have a more complicated history. Notice also the sharp rise in lift, presumably due to a rise in angle of attack soon after launch.

References

(Starting points for future study should be the Azuma and Beppu papers, together with the work of Hess and Newman. King gives some nice ideas for experiments.)

Azuma, A., G. Beppu, H. Ishikawa, K. Yasuda, Flight dynamics of the boomerang, part I: Fundamental analysis, *Journal of Guidance, Control and Dynamics* 27, 545–554, July–August 2004.

Bahn, P. G., Flight into pre-history, *Nature* 373, 562, 1987.

Bahn, P. G., Return of the Euro-boomerang, *Nature* 329, 388, 1987.

Battipede, M., Boomerang flight mechanics: Unsteady effects on motion characteristics, *Journal of Aircraft* 36 No. 4, 689–696, 1990.

Beppu, G., H. Ishikawa, A. Azuma, K. Yasuda, Flight dynamics of the boomerang, part II: Effects of initial conditions and geometrical configuration, *Journal of Guidance, Control and Dynamics* 27, 555–562, July–August 2004.

Hall, S., Boom in'rangs launches old toy into new orbit, *Smithsonian* vol. 15, pp. 118–124, 1984.

Hess, F., The aerodynamics of boomerangs, *Scientific American* 219, 124–136, 1968.

Hess, F., A returning boomerang from the Iron Age, *Antiquity* 47, 303–306, 1973.

Hunt, H., Bang up a Boomerang, Millennium Mathematics Project, University of Cambridge (http://pss.maths.org.uk/issue7/features/boomhowto/index.html) January 1999.

King, A. L., Project boomerang, *Am. J. Phys*. 43, No. 9, 770–773, 1975.

Luebbers, R. A., Ancient boomerangs discovered in South Australia, *Nature* 253, 39, 1975.

Musgrove, P., Many happy returns, *New Scientist* 61, 186–189, 24 January 1974.

Newman, B. G., Boomerangs, *Aerospace*, 13–18, December 1985.

Sharpe, J. W., The boomerang, *Philosophical Magazine* vol. 10, 60–67, 1905.

Thomas, J., Why boomerangs boomerang (and killing-sticks don't), *New Scientist*, 838–843, 22 September, 1983.

Valde-Nowak, P., A. Nadachowski, and M. Wolsan, Upper Paleolithic boomerang made of a mammoth tusk, *Nature* 329, 436–438, 1987.

Walker, G. T., On boomerangs, *Phil. Trans. Roy. Soc. London* Series A, 190, 23–41, 1897.

Books:

Ruhe, B., and E. Darnell, *Boomerang: How to Make and Catch It,* Workman Publishing, New York, 1985.

Jones, P., *Boomerang: Behind an Australian Icon,* Ten Speed Press, 1997.

12
Samaras

Figure 12.1. Sketch of the seed of a sugar maple, a familiar samara. This, like many other samaras, is a double seed.

Samara is Latin for "seed of an elm" and is a term applied to winged seeds in general. Among the many examples are the seeds of ash, maple, sycamore, and pine. These come under the purview of this book since they are spinning lifting bodies: while they resemble helicopter rotors in form and function, the distinction is that the whole body rotates.

They are autorotators—apart from possibly a spin induced at release, they are unpowered. They maintain their rotational kinetic energy from the airflow across their lifting surfaces. In addition to biological samaras, many toys exhibit autorotation—rotating kites, for example. A few origami folds can make an effective autorotator from a sheet of paper. It is also worth noting that although most of its flight relies on the initial rotation induced by the throw, the terminal phase of a returning boomerang's flight often involves autorotation and thus a boomerang could be considered a type of samara.

Finally, the samara wing architecture is employed by some modern smart munitions, and might be applied to planetary probes as an alternative to descent control by parachute.

Samaras and other architectures have recently been embraced by a fashion for "biomimetic" (i.e., biologically inspired) engineering designs: the application of samaras to possible space exploration is an example. But this is nothing new—a pioneer of aeronautics, Sir George Cayley (who among other things originated the wing-plus-tail aircraft architecture), made a close study of sycamore samaras, and the famous physicist James Clerk Maxwell studied autorotation, as did Flettner.

Investigations of Biological Samaras

Samaras, sometimes called winged or plumed seeds, or seed-wings, come in a striking variety of forms. Some are single- or double-winged, and a few are cruciform. Most autorotate in a tight spiral like a helicopter rotor, although some (the *Zanonia* a vine in the cucumber family, found in the Indonesian jungle) glide in wide lazy spirals like a circling airplane. Still others lack a weighted leading edge, and so tumble as well as spin.

Figure 12.2. The unsual gliding samara *Alsomitra macrocarpa* (some 13 cm across). The seed is in the center of the papery span, such that this samara has a minimal tendency to spin, instead gliding in a wide spiral. It achieves a glide ratio of 8:1.

The function of a samara's wing is to retard the descent of the seed from the tree's canopy, and thereby enable the seeds to be dispersed over a larger distance by wind. This provides the tree with an evolutionary advantage.

A number of studies into the flight performance of natural samaras have been conducted by botanists. Many field investigations involve simply the counting of seeds around a source tree. More detailed studies have investigated the flight behavior of individual seeds by drops in controlled conditions, sometimes with stroboscopic illumination to study their dynamics (e.g., Walker, 1981; McCutcheon, 1977; Green, 1980).

Another experimental approach is to construct samaras from conventional engineering materials (plastic, paper, etc.) and study the variation in flight performance with wing-loading, configuration, etc. Seter and Rosen (1992) and Yasuda and Azuma (1997) have adopted this approach to study aerodynamic behavior, while Augsperger and Franson (1987) measured the dispersal statistics of artificial samaras dropped from a 40 m tower.

In some clever experiments after exploring the "as built" performance of the gliding samaras, Yasuda and Azuma attached a tiny rod and weights, forcing the seed to fly at different angles of incidence. They determined that the seed glides with a fairly low lift coefficient, but at conditions intermediate between the best L/D (i.e., glide distance in still air) and minimum sink rate (maximum drift distance in wind) and concluded this may approach an optimum compromise for dispersal of the seeds. The modest lift coefficient forces a higher glide speed, which may be advantageous in penetrating through wind.

Aerodynamic Performance

Typical samaras are a couple of centimeters in span and operate at terminal velocities of a few tens of centimeters per second, giving flight Reynolds numbers of around 1000. At such low Reynolds numbers, conventional gliding flight will be poor (the only conventional glider, *Zanonia*, is perhaps the largest seed and thus has a high enough *Re*).

A bewildering array of samara shapes are found in nature (see Figure 12.3). Most typically they have the seed mass at one end, and have a ribbed structure that improves the aerodynamics via surface roughness and keeps the center of gravity forwards.

Some samaras lack this asymmetry and "tumble," i.e., spin about their spanwise axis—much like rotating kites (these are sometimes called "rolling" samaras, which is descriptive, although perhaps confusing in that strictly they are pitching!). In general the sink performance of these seeds is poorer than the helicopter type.

One might define a figure of merit for a samara—essentially a free-fall drag coefficient. Evolution's goal is to maximize the seed dispersal (via minimizing the terminal velocity) while minimizing the cost of producing that dispersal. That cost presumably relates to the mass of the lifting part of the seed, or to first order, its area. Thus we define

$$C = 2mg/(RSV^2),$$

Table 12.1. Parameters of some typical samaras from Green (1980). Azuma and Yasuda (1989) and Matlack (1987) also provide large tables of data.

Species	Mass mg	Radius cm	Disc Loading mg/cm^2	Rotor Solidity %	Terminal Velocity cm/s	Angular Rate rad/s	Tip Speed cm/s	Reynolds Number	Advance Ratio
White Ash	70.4	2.8	2.8	7.2	160	91	254	3111	1.59
Green Ash	26	1.8	2.6	8.2	162	122	216	2025	1.33
Tulip Tree	31.2	2.3	1.9	11.2	121	60	134	1933	1.11
Sugar Maple	53.4	2.3	3.3	11.4	102	149	338	1629	3.31
Box Elder	41.3	2.2	2.9	15.8	92	136	292	1406	3.17
Red Maple	14.2	1.5	2.1	15.9	66	132	191	688	2.89
Silver Maple	158.9	3.3	3.3	12.7	87	99	373	1994	4.29

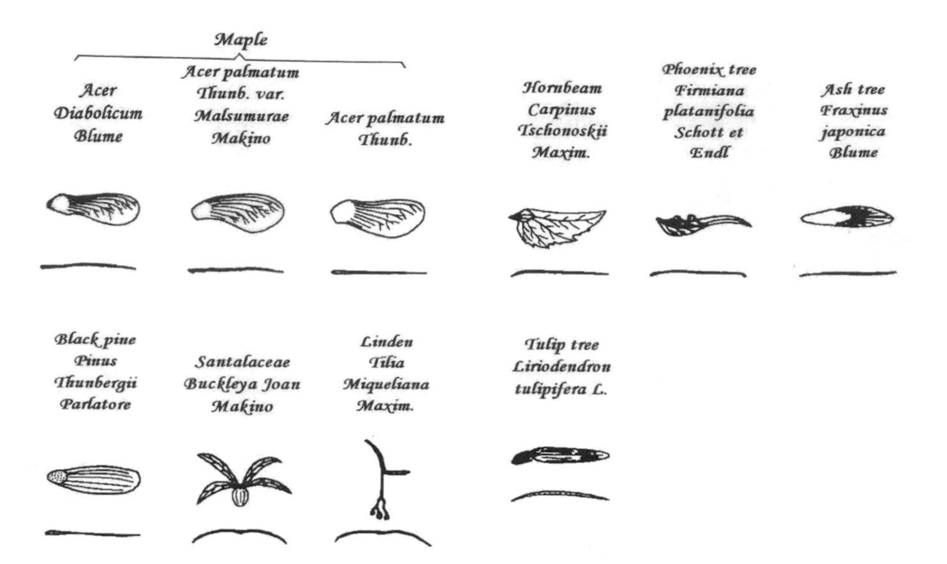

Figure 12.3. A selection of samaras (not to scale) illustrating the large range of seed-wing shape. The planform shape represents a compromise: blade area is most useful towards the tip, where the spin velocity is highest, yet losses due to tip vortices would be undesirably high were the tip chord too high and so the wing widens, then tapers.

where m is the total mass (dominated by the seed), and RS is the wing area (equals the disc area times the solidity in Table 12.1). This quantity has values of 2–6 (with larger values for the larger seeds in the table)—in other words, autorotating wings provide a much slower descent rate than the same area would yield if it were a parachute or similar decelerator (for which the drag coefficient would be ~0.5 to 1).

Azuma and Yasuda (1989) make a detailed survey of the aerodynamic characteristics of a variety of samaras, and note that the rate of descent is always between 0.2 and 0.8 times the tip speed.

Samara Stability

The descent speed, rotation rate, pitch of the wing, and the cone angle are all coupled, and their various interactions in general lead to stable flight. Norberg (1973) conducts a detailed analysis of the dynamics of samaras, drawing analogies with helicopter and bird flight. His analysis also takes side-slip into account, although in most cases this is not an important effect. The interaction of side-slip with the samaras rotation gives many samaras a slightly helical descent path.

Two points to note are that the spanwise area distribution is nonuniform, causing the lift force to be biased outboard (according to Norberg's analysis of the *Acer platanoides* (Norway maple) samara, half of the lift comes from the outer third of the wing). This increases the moment arm of the lift. The other significant element is the ratio of the mass of the seed to the wing—in maple seeds the wing only makes up 15% of the total mass. The lower the seed mass, the less asymmetric the mass distribution and the weaker the tendency for rotation to flatten the cone angle. Some samaras, such as the *Tachigalia versicolor* studied by Augsperger and Franson (1987), have the seed at the center of the span, and these seeds tend to tumble rather than spin.

The cone described by the motion of the samara blade is defined by the balance between the lift force on the blade (which tends to narrow the cone) and the centrifugal force which tends to cause it to describe a horizontal circle. As can be easily demonstrated with a circle of paper made into a cone, the cone angle determines the compromise between drag efficiency (higher for a flatter cone) and stability (higher for a sharp cone).

The pitch stability depends on the relative positions of the center of mass and center of pressure. For most samaras the center of mass is about 30% of the chord in from the leading edge. The center of pressure depends on the angle of attack—near the seed the angle of attack is high and the center of pressure is halfway across the chord, leading to a nose-down pitching moment. However, towards the tip where most of the lift is generated, the angle of attack is lower and the center of pressure is around the quarter-chord point, leading to a nose-up pitch moment. Integrating along the span of the samara, an equilibrium is found, leading to a stable pitch angle.

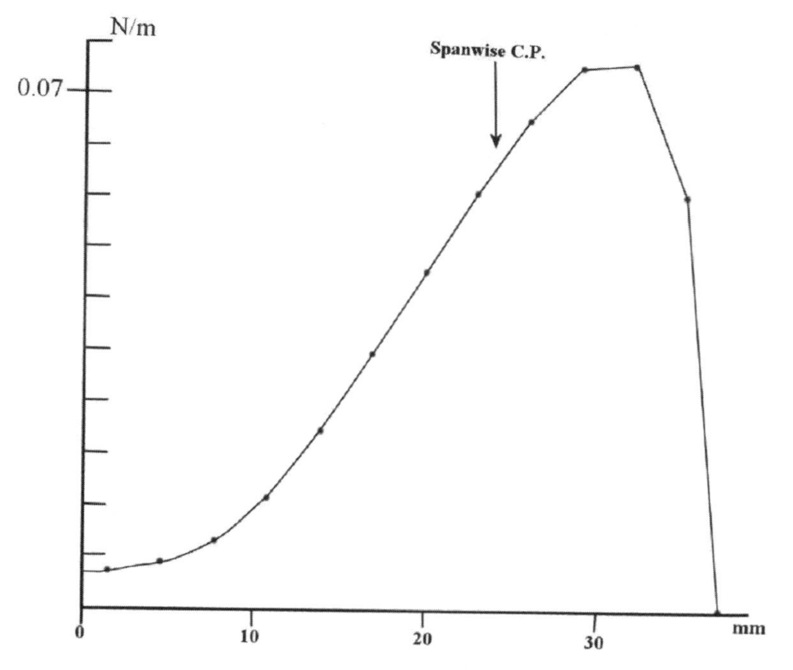

Figure 12.4. Norberg measured and inferred the weight and lift distributions across a samara. Due to the rotational component of relative wind over the wing, the center of lift is rather outboard of the center of mass.

Flight Performance of Model Samaras

Yasuda and Azuma (1997) made a study of the flight characteristics of a suite of model samaras, together with some natural ones, and natural but modified examples. These were introduced into a vertical wind tunnel ($30\,cm^2$) and their flight characteristics determined with stroboscopic photography.

Natural samaras had their center of mass moved by means of small balance weights, and the positions which permitted autorotation and thus stable, slow descent could be determined. The same exercise was undertaken for various shapes of wing made from thin styrene foam, of 0.4 mm thickness and typically $8 \times 3\,cm^2$ size. These wings had a mass of about 0.12 g, to which a weight of similar mass was attached.

Yasuda and Azuma found that plane wings with unmodified surface roughness were in fact very difficult to get to autorotate. In the small range of center of gravity positions that did rotate, the flight performance was quite poor—a terminal velocity of about 1.00 m/s, a rotation rate of only 350 rpm and a large coning angle, some 55 degrees. The center of gravity positions that worked, surprisingly, had the center of gravity close to the corner of the wing root and trailing edge. These parameters should be compared with real samaras (0.9 m/s) with a 1000 rpm spin with 10 degree coning angle.

Yasuda and Azuma found that several modifications to their model samaras improved their performance. All of these modifications are exhibited by natural samaras, suggesting that the natural examples are quite well optimized by evolution.

First, the wing should have a negative camber, at least at the inboard portion (but recall it is the tip of the wing that contributes most to lift). In other words, it should be bent convex downwards—this causes the aerofoil to settle at a stable and useful angle of attack. Second, the leading edge of the wing should be thickened—this is achieved in nature (and in the models) by the addition of ribs (in the model case, these were 0.6 mm glass fiber rods glued onto the wing). These ribs also appear to enhance spin performance by modifying the surface roughness—natural samaras with their ribs filed down performed much more poorly.

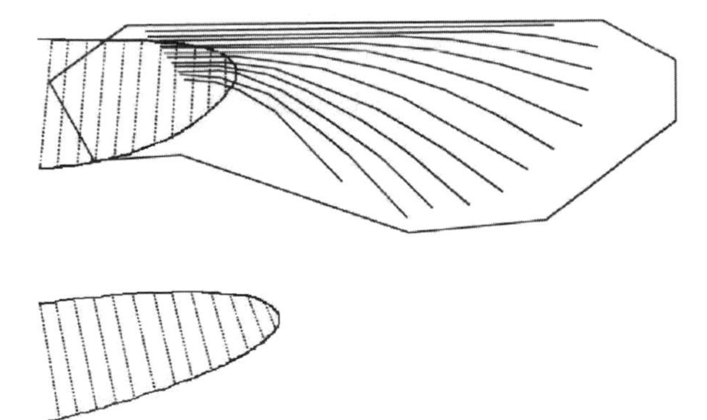

Figure 12.5. Permitted center of gravity location for an artificial samara (see Yasuda and Azuma, 1997). The two shaded regions denote the center of gravity location for which autorotation reliably occurs: the lower region is where the center of gravity needs to be for a flat unribbed samara; the ribbed structure displaces the permissible center of gravity locations to the upper shaded zone, obviously more conveniently achieved.

Micrometeorology and Release Parameters

Simple experiments show (Martone, 2001) that a typical samara must fall a meter or so in order to spin up. Further, if the outboard tip of the samara is trimmed off, the seed takes longer to begin autorotation, and if enough of the outboard part is removed, the seed may never fly at all, simply falling vertically while spinning around that axis.

In still air, most samaras fall near-vertically, their spiral having a rather small radius. However, in reality wind will cause the seeds to disperse laterally. If the tree is modeled as a point source of seeds at some elevation above the ground, the seeds are deposited with radial distribution peaking at some value (which may be a nonzero distance from the source) and tailing off at long distances—a skewed distribution such as a Weibull distribution or a log-normal distribution appears to fit the data quite well. If the winds are in a prevailing direction, then this distribution will be smeared in that direction as a plume or streak. The presence of other trees of course modifies the windspeeds experienced by the seeds.

Another factor is the presence of vertical winds. Even a transient updraft associated with an eddy can have a dramatic effect on the distance traveled by a given seed. Note that most simple meteorological stations do not record the vertical component of wind, so data on this effect are sparse. One field observation (Augsperger and Franson, 1987) did note some 25–30 samaras lofted by an updraught from a tropical tree *Tachigalia versicolor* and carried some 172–277 m (the closest of these being some 65 m beyond the normally released samaras).

An important factor in the dispersal distance is when seeds are released. As yet there is no consensus on circumstances of release; some data suggest a preferential release during periods of low humidity (typically early afternoon, when winds are strongest). Strong winds may directly cause the detachment of seeds.

Were all seeds to be released in still conditions, the dispersal would obviously be minimal, due only to the stochastic gliding distance. Seeds released at purely random (i.e., in the long term, uniformly with time) will have a distribution of dispersal distance that indicates the variability of the wind. However, if seeds are only released when the wind is strongest, the dispersal distance will be biased towards higher values, which is the preferred circumstance for the tree.

As a last thought, note that although tumbling samaras like the tulip ash may have poorer *flight* performance than conventional rotor samaras, flight parameters may not be the ultimate determinant of dispersal performance. It may be that tumbling samaras are more easily moved across the ground by wind after the samara has landed.

Samara Munition: Smart "Skeet"

We move now to artificial samaras. As discussed in an earlier chapter, smart weapons for attacking dispersed targets such as armored formations can exploit spin to scan sensors, and the footprint of a warhead, over a wide area without needing to apply destructive force to the whole area. Such munitions are usually in the form of short cylinders which rotate in a near-horizontal plane. One such warhead, the Textron Sensor Fuzed Weapon, is referred to as "Skeet"™, after the clay catapult-launched cylindrical projectiles used as targets for shooting (see chapter 10).

Developments in the late 1970s found that a samara wing inspired by the maple seed offered good performance by extending the descent time of the warhead, but in a more compact package than could be achieved with a parachute. Textron's Selectively Targeted Skeet (STS), a derivative of the Sensor Fuzed Weapon (SFW), is one result.

The SFW is intended principally for air attack of armored formations and detonates in the air. Its warhead is shaped, with a copper sheet facing downward, such that a rod of metal (in this case, copper) shoots down (an "explosively formed penetrator," EFP) to attack a target such as a tank on its top surface, which is usually only lightly armored. The detonation is triggered by laser and/or infrared sensors which are boresighted with the EFP axis—the combination of the canted axis swept around by the munition's spin and its horizontal motion across the ground means the potentially lethal area is very large. Rather than inefficiently using a large explosive to devastate an entire area, this "smart" weapon covers this area but effects its destruction only on the target within this area.

The BLU-108 munition is an air-dropped canister which deploys four of these "skeet" warheads. These are shaped somewhat like a hockey puck, around 13 cm in diameter with a weight of 3.4 kg. The BLU-108 releases a drogue parachute, which decelerates it and holds it vertical. At a set altitude the parachute is cut free and a rocket motor spins the canister and propels it upward. The four skeets are flung out horizontally, in four directions, thereby covering an area some 260 m across. Each spinning and coning skeet follows a flat parabolic trajectory, making a spiral scan pattern on the ground that is widest at the highest point of the trajectory (i.e., apogee; see Figure 12.6).

A single F-16 fighter can carry four SFW dispensers which each release ten BLU-108s, for a total of 160 submunitions; an F-15 can carry 10 canisters. Release and drogue parachute timing is controlled to optimize dispersal, depending on flight speed. This weapon was used with considerable success in the Gulf War.

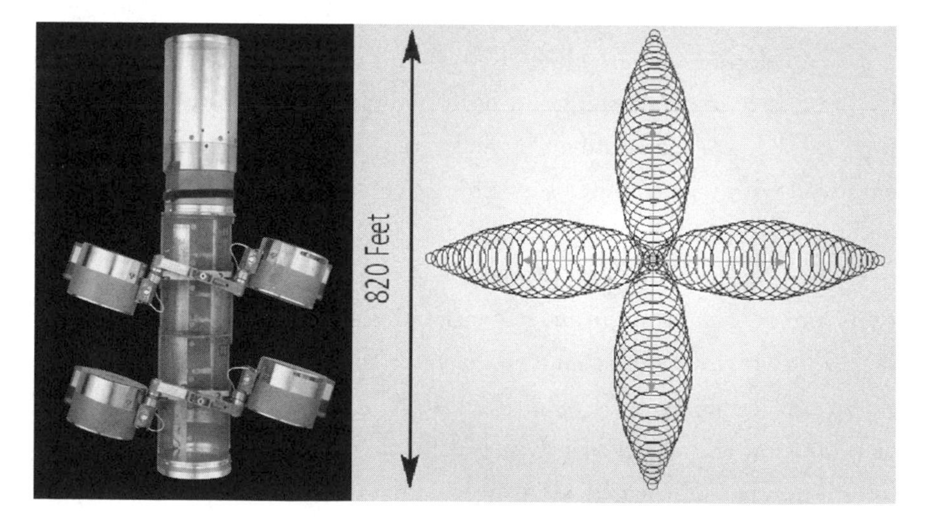

Figure 12.6. BLU-108 munition throws out four independent sensor-fuzed warheads which use their spin to scan a pattern on the ground. The combination of the four munitions launched orthogonally describes a cruciform joint footprint: the tips of the four arms become narrow as the submunition altitude decreases. Images courtesy of Textron Systems.

STS Samara Munition Development

The SFW is designed as an area weapon, engaging multiple targets within an area. While this method is invaluable for the effective defeat of target arrays such as an armored column on a road, the large area can prevent the use of the weapon in areas where collateral damage is prohibited, as in many post–Cold War conflicts. An example would be a surface-to-air missile placed next to a historic edifice or a hospital. For this reason, the Selectively Targeted Skeet submunition was developed. It uses the SFW warhead, but employs a samara wing to slow its vertical descent to generate a scan 30 degrees from vertical. The STS can be individually dispensed from rockets, unmanned air vehicles, or aircraft to a selected target, one submunition at a time. Its search pattern (shown in Figure 12.7) is a collapsing spiral that ensures target detection and defeat. The STS is spun to near its natural scan rate upon release. This is done by either the dispenser or by a small rocket motor within the submunition.

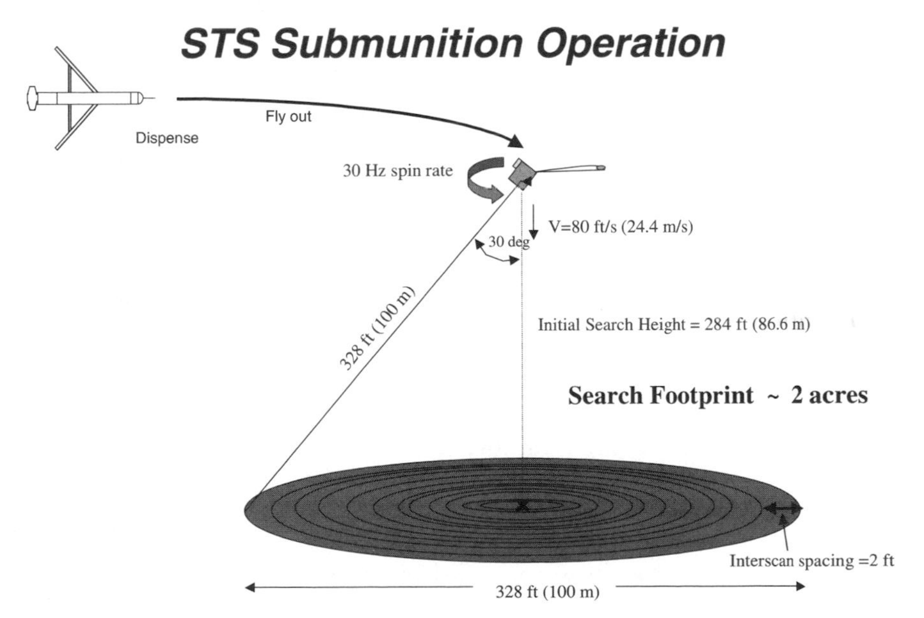

Figure 12.7. The operation of the STS munition. After release, it scans a wide area using a samara wing to control the descent speed and rotation. Figure courtesy of Textron systems.

The wing of the 3.6 kg STS is about 25 cm long and around 8 cm across. The STS nominally begins its search from 100 m altitude. Its sensors sweep at a nominal cone angle of 30 degrees, driven by the autorotation of the wing. (The search altitude is driven by the lethal range of the warhead—the munition can in principle be released from higher up).

Figure 12.8. Selectively Targeted Skeet munition. The samara wing with tip mass is mounted on the opposite side from the sensor package, which uses infrared light and other methods to identify a target. The munition detonates in midair, causing the dished copper liner to form a hypervelocity slug of metal which penetrates the soft upper skin of the target. Photo courtesy of Textron Systems.

The parameters used in the STS are of course not randomly chosen. Early wind tunnel tests and modeling were performed in the 1980s with a variety of configurations to explore how the scan angle, spin rate, and sink speed were affected by the wing size and the tip mass and its location.

Simulations by Kline and Koenig (1984) suggested that the maximum practical scan angle achievable for the representative munition size and shape was about 50 degrees—this version had a long wing (which they called a fin) with as large a mass as possible on its end. Tip mass (from .042 to 0.26 lb.) and fin length were varied from 6 cm to 20 cm. The lowest values of these quantities gave a spin angle of only 10 degrees. The angles were reduced for a heavier munition.

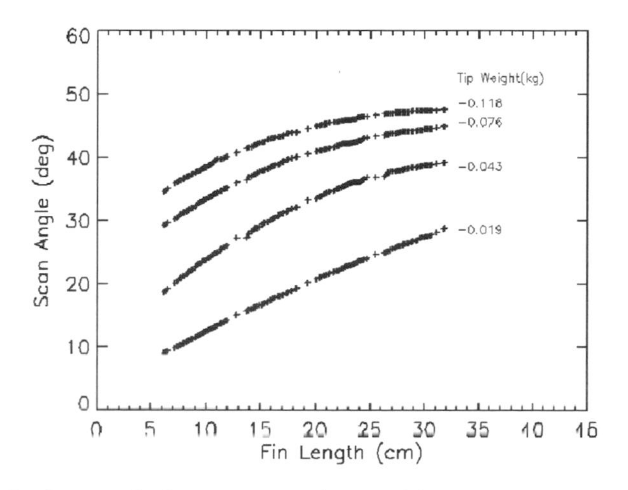

Figure 12.9. Scan angle for a samara-wing munition. Increasing the tip mass and the length of the fin (span of the wing) increases the scan angle. Both variations also affect the descent rate.

After the simulations (using 6 degrees of freedom—the wing was essentially considered as rigid), models were tested in a vertical wind tunnel. The model was a 2.78 lb. cylinder (4.75 in. diameter 3.4 in. long, made of Lexan—35% of mass of real munition) with 0.085 lb. tip mass on 7.5 in. span, 3 in. chord wing of doubled 3 oz/yd.^{2} nylon.

After release at 4 revolutions per second, the unit spun up to 7 revolutions per second within 2 seconds: its terminal velocity was 77 ft./s. Notably, the flexible wing was curved in steady-state flight, tip to mount inclined at about 35 degrees to the top of the munition, but with an upwards curvature radius of 30 cm.

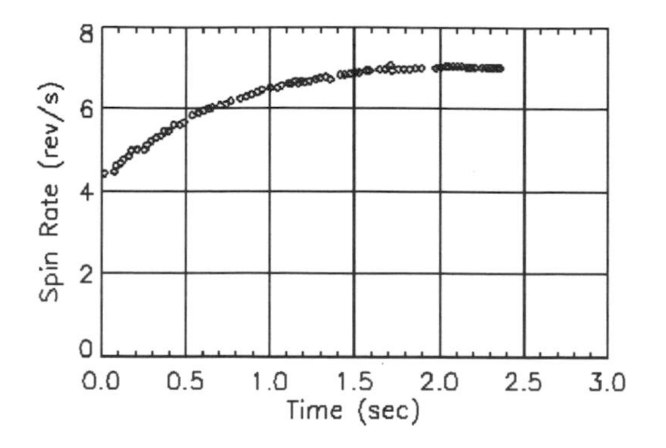

Figure 12.10. Spin-up of a samara munition model released into the airstream in a vertical wind tunnel.

The tests found that the fin was more stable, especially during spin-up, if the tip mass was mounted such that its center of gravity was at the quarter-chord position on the fin rather than at the half-chord position.

A more sophisticated model was developed by Crimi (1984, 1988), who also included a treatment of the aeroelastic effects on the wing. He considered the tip mass as if it were linked by four elastic cords to the center body—these fictitious elastic cords acted to capture the net behavior of the wing material. The resulting system had a total of 11 degrees of freedom.

The lift and drag characteristics of the wing as a function of angle of attack were adapted from those of the NACA 0012 airfoil section. Interestingly, since the samara wing has a sharp leading edge, the coefficients used were those of the NACA 0012 shifted in by 180 degrees—in essence the simulations assumed a NACA 0012 section flying backwards! We are reminded in this approach of the backwards flying characteristics of this section, used in the design of the Aerobie flying ring (chapter 9), where a section's backwards flying characteristics are important.

Crimi (1988) notes an important point in testing (though he uses the observation to excuse poor agreement of models with tests in a couple of cases). Since the models were hand-launched into the wind tunnel at low spin, they took some time to spin up. However, since the wind tunnel speed must be adjusted to match the descent speed and it takes a finite response time to adjust, it was difficult to achieve steady-state conditions.

In 2000 through 2003 Textron extended the work done by Crimi et al. into the design of the STS submunition. The analytical models were refined and a series of gas gun launches and helicopter drop tests were conducted to validate the models and arrive at a reliable samara wing design. A wing and its deployment system were developed and demonstrated to provide reliable function at initial submunition horizontal velocities from stationary to over 135 m/sec: tracking camera images of the munition in autorotating flight are blurred due to the high (30 Hz) rotation rate, but confirm the scan angle and descent rate of 30 m/s.

Rigid-Wing Samara Munitions

An alternative approach to the same goal as STS was taken by the BONUS munition developed by the Swedish firm Bofors with GIAT of France. This 138 mm diameter munition is launched in a 155 mm artillery shell (like the SADARM), but uses a rigid wing that is deployed in flight to achieve a lower descent rate and set up a spin.

Released at an altitude of 175 m, it descends at 45 m/s with a spin rate of 15 revolutions per second. During its short descent, the 6.5 kg munition, of which two are carried per shell, sweeps out an area of some 32,000 m^2. If the munition, with 3 multiband infrared sensors and an altimeter, detects a target, its explosively formed projectile warhead (a tantalum sheet, rather than copper) can penetrate up to 130 mm of steel armor.

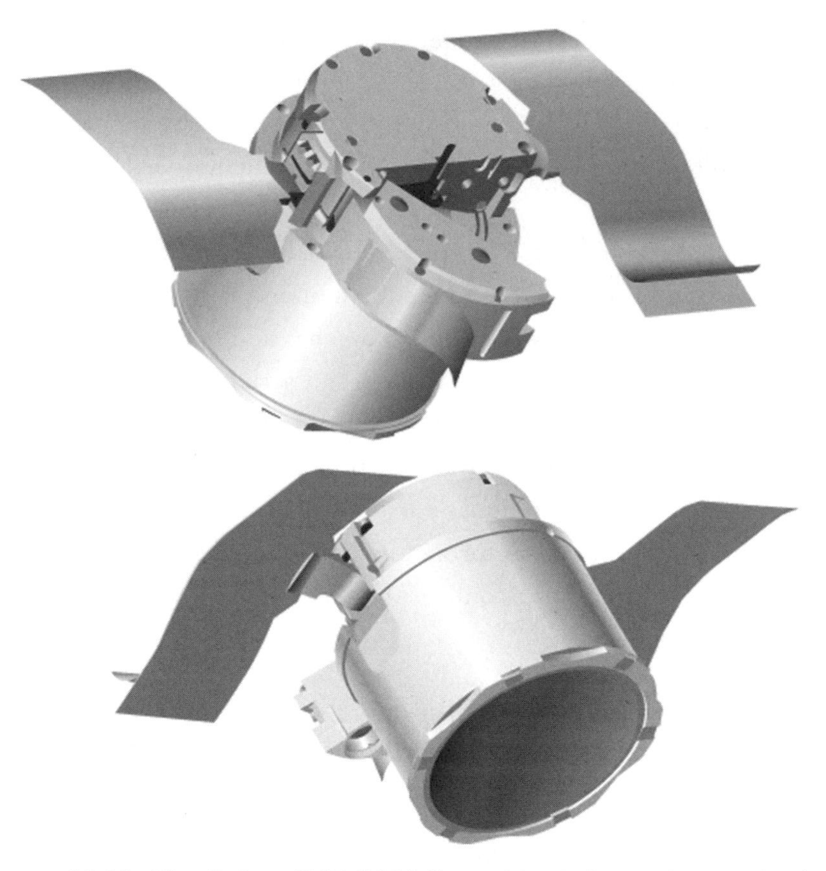

Figure 12.11. The Bofors–GIAT BONUS munition. Wing surfaces spring into position to set the scan angle and spin rate during descent. Image courtesy of Bofors.

Rocket-Propelled Guided Samara

Samara munitions, as their rotating symmetry would suggest, tend not to have a net horizontal drift relative to the air mass in which they are descending. If that air mass is moving due to a background horizontal wind, then the munition will move accordingly. To correct for wind drift, or perhaps to expand the lethal footprint of the munition, the possibility of guiding the samara has been explored. It might be that the munition can sense targets that are beyond the range of its warhead, and thus by moving horizontally the target can be engaged.

However, the samara wing, by virtue of its inexpensive simplicity, cannot be used for this sort of diversion. Instead, a sideways-pointing rocket motor is used to "kick" the munition (Pillasch and Pangburn, 1993). Because it is spinning rapidly, the rocket burn must be short in duration so that the thrust is exerted over a modest arc of rotation, leading to a large net impulse in one direction (i.e., the impulse must be short in order that the spin stabilization effect exploited in conventional rocket vehicles doesn't work!). In practice, this means the burn must be 8 ms or less.

This concept was demonstrated with drop tests from a helicopter. A spin drop fixture used an electric drill motor. With the helicopter hovering at the desired drop altitude, the unit was spun up to 3 Hz to check that the samara wing was deployed, then spun to 10 Hz for release. The motor firing 12 s after release, with the 13 kg unit having spun up to 17 Hz and in a steady descent at 10 m/s, caused an impulsive change in horizontal velocity. The impulse, at an altitude of 300 m, caused the impact point to be displaced some 69 m from where it was predicted in the absence of thrust. The impulse, as might be expected, caused a change in the body axis orientation, but did not affect the spin rate.

Future Applications

Recently, the notion of using a samara configuration for the deployment of instrument packages, perhaps into planetary atmospheres, has been explored (Thakoor and Miralles, 2002—in fact the idea is rather older; Burke 1988). A flat wing shape might be integrated with a solar cell and antenna, in a robust and compact configuration. This is a much simpler architecture than a more conventional parachute-borne package which would require some sort of deployment mechanism. An additional advantage for atmospheric science investigations is that such a configuration has a clear view of the sky, permitting measurements of light scattering, absorption of sunlight by gases, etc.

In addition to exploring other planetary atmospheres, perhaps released from balloons, such platforms might (if adequately stable against strong turbulence) be useful in exploring environments on Earth such as tornados.

References

Augspurger, C. K., and S. E. Franson, Wind dispersal of artificial fruits varying in mass, area and morphology, *Ecology* 68 (1), 27–42, 1987.

Azuma, A., and Y. Okuno, Flight of a samara, *Alsomitra macrocarpa. Theor. Biol.* 129: 263–274, 1987.

Burke, J. D., Atmospheric autorotating imaging device, NASA Patent NPO-17390, 1989.

Crimi, P., Analytical modeling of a samara-wing decelerator, AIAA Paper 86-2439, October 1986.

Crimi, P., Analysis of samara-wing decelerator steady-state characteristics, *Journal of Aircraft* Vol. 25 41–47, January 1988.

Greene, D. F., and E. A. Johnson, A model of wind dispersal of winged or plumed seeds, *Ecology* 70 (2), 339–347, 1989.

Green, D. S., The terminal velocity and dispersal of spinning samaras, *American Journal of Botany* 67 (8), 1218–1224, 1980.

Horn, H. S., R. Nathan, and S. R. Kapln, Long-distance dispersal of tree seeds by wind, *Ecological Research* 16, 877–885, 2001.

Kline, R., and W. Koenig, Samara type decelerators, AIAA Paper 84-0807, April 1984.

Martone, R. L., The flight of winged seeds, www.sas.org/E-Bulletin/2002-11-08/features/body.html, 8 November 2002.

Matlack, G. R., Diaspore size, shape and fall behavior in wind-dispersed plant species, *American Journal of Botany* 74, 1150–1160, 1987.

McCutchen, C. W., The spinning rotation of ash and tulip tree samaras, *Science* 197 (4304), 691–692, 1977.

Norberg, R. A., Autorotation, self-stability and structure of single-winged fruits and seeds (samaras) with comparative remarks on animal flight, *Biological Reviews* 48, 561–596, 1973.

Pillasch, D., and D. Pangburn, Impulse guided samara decelerator, AIAA 93-1234, *12th RAeS/AIAA Aerodynamic Decelerator Systems Technology Conference*, London, May 10–13, 1993.

Seter, D., and A. Rosen, Study of the vertical autorotation of a single winged samara, *Biological Review* 67, 175–192, 1992.

Thakoor, S., and C. Miralles, Seed-wing flyers for exploration, *(NPO-21142) NASA Tech Briefs* Vol. 26, No. 1, page 47, January 2002.

Walker, J., The amateur scientist: The aerodynamics of the samara: winged seed of the maple, the ash and other trees, *Scientific American* 245, 226–238, 1981.

Yasuda, K., and A. Azuma, The autorotation boundary in the flight of samaras, *Journal of Theoretical Biology* 185, 313–320, 1997.

http://www.boforsdefence.com/
http://www.systems.textron.com/

13
Skipping Stones and Bouncing Bombs

One of childhood's pleasures, and one I still enjoy, is skipping stones. There is the pleasing aesthetic of ripples spreading radially from the contact points and the challenge of apparently defying gravity. To supplement the remarkable variety in the dynamics, there is instant gratification both audibly and visibly—within a second a churning plop will tell you you've thrown it wrong.

Despite the charm and ubiquity of this practice, the phenomenon is not terribly well understood. The problem involves nonsteady hydrodynamics at a fluid interface, together with the aerodynamics and gyrodynamics that are familiar to us elsewhere in this book.

Although the easiest and most common incarnation of stone-skipping is of a rather flat stone spinning in a near-horizontal plane (like a small, heavy Frisbee), we should acknowledge that it is perfectly possible to skip near-spherical stones, and indeed skipping cannonballs was a recognized technique of early naval artillery (e.g., Johnson, 1998a). It was only in connection with an unusual cylindrical skipper, the Bouncing Bomb of World War II, that any quantitative attention was paid to the problem. Spin plays an important role here, although in a quite different way from the skipping stone.

Although the motivation for including skipping stones in this book is geared mostly to their spinning nature, the discussion of impulsive contact with a liquid surface also prompts mention of two related instances: the splashdown of a space capsule, and the splashing gait of the basilisk lizard, an animal that literally walks on water. It is hoped the reader will forgive these nonspinning digressions.

The Bouncing Bomb

The "Bouncing Bomb" was developed by the British engineer Barnes Wallis during World War II, in order to breach the Möhne dam in the industrial Ruhr valley (Wallis had previously been an accomplished designer of airships, rejecting the cigar shape common in the early years in favor of an ellipsoidal shape, which had much better performance). The Ruhr dams were recognized by Admiralty studies even in 1938 as being vital to the German industrial effort, both in assuring the water supply to the factories and population, and for the hydroelectric power they generated. To breach a heavily defended narrow dam was impossible by conventional bombing, which would have needed impracticably many hits to breach the dam. Nets suspended from floating booms in front of the dam prevented attack by torpedo. Wallis developed a concept of a very large Earth-penetrating bomb (see Chapter 4), but the aircraft to drop such a large weapon were not available early in the war.

To initial incredulity, Wallis developed a scheme whereby a spinning spherical or cylindrical bomb would be dropped onto the reservoir's surface by a low-flying bomber aircraft such that it bounced several times, skipping over the torpedo nets and impacting the dam near the waterline.

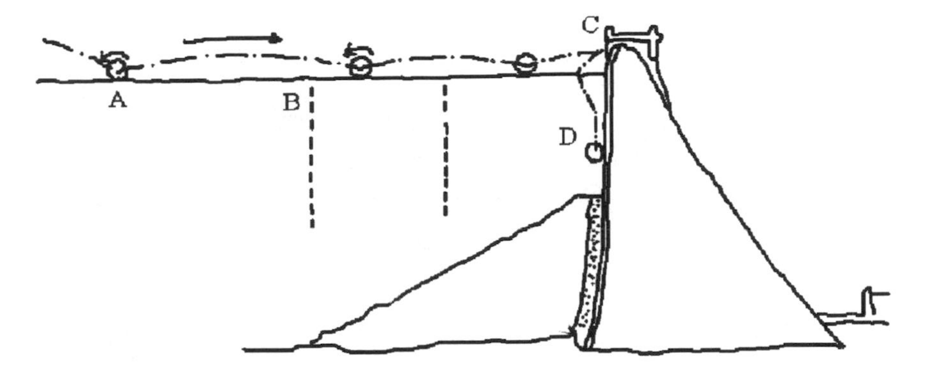

Figure 13.1. Schematic of the Bouncing Bomb principle of operation. The bomb is dropped at a shallow angle onto the reservoir surface with a backspin (A). This permits it to skip across the surface of the reservoir, evading the torpedo nets (B). After impact with the dam at (C), the bomb rolls down the front of the dam, with the spin helping to keep it "stuck" to the dam wall. A pressure switch causes the bomb to detonate at the dam wall at an optimum depth (D).

After scale-model experiments at Nant-y-Gro in Wales to determine the amount of explosive needed to breach the dam and the optimum detonation depth, full-scale tests were performed which showed that if the release parameters of the spinning bomb were sufficiently controlled, this unusual delivery approach had a reasonable chance of success. The spin of the bomb could be controlled by a special motor installed in the bomb bay, and the forward release speed could be set simply by the airspeed of the four-engined Lancaster bomber. The vertical speed of the bomb at its first bounce was a crucial parameter, and would be fixed by the release altitude of the bomb. The formidable air defences at the dam mandated a low-level night attack and so to determine the release altitude, two searchlights were mounted on the aircraft, angled such that the two beams converged underneath the plane at the required distance (60 ft.). Flying at the desired speed (220 mph) at an altitude where the beams formed a single spot of light on the reservoir surface, the bomb would be released at a predetermined distance from the dam (this latter parameter being determined by the horizontal angle subtended by the dam—the "bombsight" was simply a piece of wood with two nails).

Figure 13.2. A Lancaster drops a bomb in this still from cine footage during trials. The bombrack and chain are just visible above the bomb. Note the low altitude of release. IWM FLM2300—Imperial War Museum Archives, used with permission.

Figure 13.3. The bomb skips with a splash (this is in fact the second or third bounce. Even though this was a dummy bomb for trials, the proximity of the observers is impressive. IWM FLM 2343—Imperial War Museum Archives, used with permission.

Although a nonspinning bomb would skip once, spin was needed to prevent the bomb from tumbling. In fact the bomb (code-named "Upkeep") was given a 500 rpm backspin by a motor underneath the bomber—spin in either direction would do for stability, but giving it a backspin helped the bomb roll down the front edge of the dam to the optimum depth of 30 ft., where a pressure trigger detonated the weapon.

Figure 13.4. A view of the frame holding the Upkeep bomb under the Lancaster bomber. A chain can be seen between the bomb and a sprocket in order to spin the bomb up in flight. The triangular support calipers swung outwards to release the weapon. Crown Copyright.

In fact, Wallis was initially reluctant to invoke the complication of spin, preferring a spherical geometry which would not need stabilization (the weapon was called a "spherical torpedo" in the earliest designs). However, after prompting by colleagues, spin's advantages became obvious. In addition to the attitude stability, spin had two other effects—backspin provided lift via the Robins–Magnus effect and thus allowed the bomb to hit the water at a more shallow angle than a ballistic trajectory would allow. It furthermore expanded the range of impact angles over which the bomb would skip by providing hydrodynamic lift at contact.

Experiments have shown that a nonspinning sphere will skip for impacts shallower than a critical angle $\Theta_c \sim 18°/(\rho/\rho_l)^{0.5}$, where (ρ/ρ_l) is the density of the weapon relative to the liquid (i.e., its specific gravity). The eventual design of the bomb was a steel cylinder—essentially a depth charge—with a 50 in. diameter, 60 in. long, weighing 9250 lb. and containing 6600 lb. of RDX explosive. With a specific gravity of 2.17, the skip angle without spin would therefore have been only 12° (Johnson, 1998b). The 52 rad/s backspin, giving an advance ratio of about 0.3, increased the critical angle to a more forgiving 16°. The bomb was released at 425 yards from the dam wall, a distance it traversed in about 4 seconds. A set of three hydrostatic pistols detonated the charge when it had sunk to 30 ft.

In the attack, made famous—if not entirely accurately—in the 1952 *Dambusters* film, the Möhne dam was breached at the 4th attempt, releasing 134 million gallons of water. The bombers went on to further targets, breaching the Eder dam. However, only 11 of the 19 bombers returned safely to base—a not untypical attrition rate at the time. (Incidentally, the "trench" scene in the original *Star Wars* film owes much to the portrayal of Operation Chastise in the *Dambusters* film.) A detailed history of the raid is given in the book by Sweetman (2002); a good short technical account is given in Hutchings (1978). Although, as here, Barnes Wallis is often the one name associated with the project, it should be noted that major contributions were made by several other scientists and engineers, not least in the exhaustive scale model tests involved in the project.

A second type of skipping bomb, named "Highball," was developed to attack ships. This bomb was to be carried by Mosquito aircraft (in turn celebrated in the film *617 Squadron*). In fact, the Navy had wished the Upkeep raid on the dams (code-named Operation Chastise) to be delayed, so that the technique could be kept secret until its application at sea.

German scientists recovered one Upkeep weapon from one of the Lancasters that was shot down (although the bomb had a 90 s time fuze as a backup to the hydrostatic pistols, since the bomb was not released, the fuzes were not armed), and quickly understood its technique of operation, recognizing it as essentially a spinning depth charge.

An 80 cm diameter, 700 kg antishipping bomb code-named Kurt was then developed in Germany. This bomb had a prismatic shape and appeared to offer better skipping performance than the scaled-down Upkeep. However, it had twice the density and therefore needed higher speed or lower release altitude to skip acceptably, and thus a rocket motor was added to the bomb. Although it could skip up to 2 km, its accuracy was poor.

Much of the technical work on bouncing bombs only came to light in the 1970s, after the 30-year secrecy period (e.g., see the work by Hutchings, 1976, and Soliman et al., 1976). In reality, of course, the

Figure 13.5. The author with a practice Upkeep weapon. The wheel of the Lancaster bomber is just to the left.

bouncing bomb was of limited application and the development of guided missiles soon after the war rapidly offered a better solution for antishipping strikes. As an unconventional engineering solution to the particular problem of attacking the dams, the bouncing bomb has a certain elegance and makes a great story.

Stone-Skipping

The practice of stone-skipping is doubtless prehistoric and is documented by the ancient Greeks. The game acquired the name "Ducks and Drakes" in medieval England (one reference dates to 1583). The skip after the first impact is called a "duck," after the second a "drake," and so on. The game may have been played with oyster shells as well as stones, and there are at least apocryphal suggestions that it may have

Figure 13.6. A stone caught in midair just after its second skip. The height of the ~ 4 cm stone can be judged from the distance between the stone and its reflection in the water surface. Note the central resurge and two curved jets of water in the second splash—the *V*-shaped spatter pattern from the first skip can be seen radiating from the circular ripple around the impact point. Photo by Dr. Elizabeth Turtle, used with permission.

been played with coins such as a sovereign. The expression "to play ducks and drakes" may have in this way come to mean "to squander," as well as "idle play." Yet there is rich physics at work.

Distance is inconvenient to measure, so performance is usually gauged by the number of skips. A long-standing world record was held by Jerdone ("Jerry") Mcghee, who threw a stone with 38 skips on the Blanco river in Texas in 1994. This record was recently broken by Rob Steiner, who accomplished 40 skips in 2002. Even these record-breaking throws last only a few seconds, and the number of skips were verified by frame-by-frame examination of a video record from an overhead camera (on a bridge above the river).

Perhaps surprisingly (though perhaps not, if the reader has followed the publication dates of most of the references on other spinning topics in this book) it is only recently that serious scientific study has been applied to the phenomenon.

Some of the analytical building blocks are quite old—Lord Rayleigh in 1876 wrote in a paper on the resistance of fluids that an elongated blade held in a horizontal stream at an angle β has a mean pressure of $\sim (\pi/5)\cos\beta\, \rho_1 V^2$ (this relation was used, for lack of anything better, in deducing the effect of spin on the bouncing bomb).

For the most part, the aerodynamic forces on a skipping stone can be neglected (though later we discuss one circumstance where this is not the case). A typical stone might have a diameter of 4 cm and a thickness of 1 cm. Its mass will therefore be of the order of 100–200 g and its weight thus a Newton or two. The aerodynamic lift on this object, flying at say 5 m/s, is of the order of 0.02 N—fifty times smaller. Thus, to a reasonable approximation, the travel of the stone through the air is ballistic. Its interaction with the surface of the water, however, is altogether different. All else being equal (it is not!), the lift and drag through water would be larger by the ratio of the densities of air and water, about 800, and thus the hydrodynamic forces would exceed weight by 800/50, or a factor of about 16.

At each skip, the stone is reflected upwards. Its downward velocity is almost completely reversed, its horizontal velocity will be reduced somewhat, and since it will typically breach the water surface trailing side first, a pitch-down moment will occur.

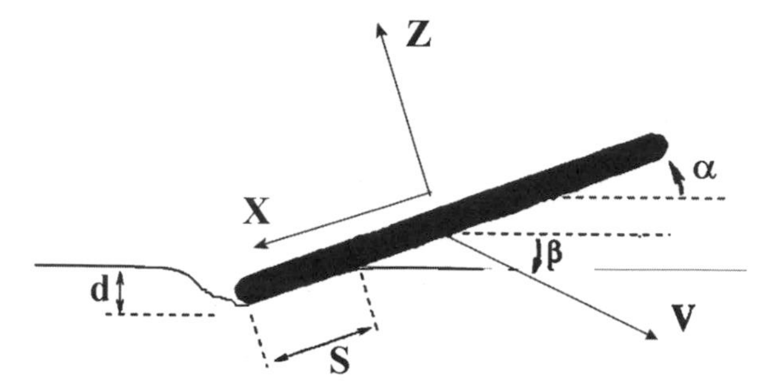

Figure 13.7. Geometry of a stone hitting the water. The stone has penetrated the water surface by a distance *d*; the *X*-*Y* plane of the stone makes an angle α with the undisturbed (horizontal) surface of the water, while β defines the flight path angle. $(\alpha + \beta)$ defines an angle of attack. *S* denotes a wetted area of the stone.

The effects of the pitch-down moment on the subsequent attitude of the stone can be minimized by giving the stone momentum bias, i.e., a high spin to give gyroscopic stiffness. As discussed later, the pitch-down moment causes the stone to precess, giving the stone a usually right-curving path.

In some cases, the extent of the stone's flight is limited by its forward velocity—as each bounce slows it down, it eventually impacts the water surface with a steep enough flight path angle so that it penetrates rather than skips. In other cases, the vertical velocity (in turn often a function of the height from which the stone is thrown) is the limiting factor. Here, the stone fails to break contact with the water during a bounce, and instead ploughs forward at the water surface, quickly slowing down as a result.

Sometimes, instead of ploughing to a halt on the water surface, the flight may be curtailed by either the impact attitude—if the stone has pitched forward, or has rolled onto its side, it will penetrate rather than skip. A rough water surface, due to wind-driven waves, makes it difficult to skip stones: even gentle undulations can magnify trajectory or attitude disturbances, such that—like a recursive calculation or other chaotic system—after several skips the impact cannot be predicted and usually results in a terminal splash.

Previous Work on Stone-Skipping

The practice is universal, and the criteria for success are intuitively obvious: a flight both long in time and distance, with many skips. Since the activity is rarely conducted with distance-measuring equipment, the most obvious metric is the number of skips, the present world record being some 40 skips.

Stong (1968) showed how stroboscopic photography could "freeze" the motion of a skipping stone, although he presented little in the way of results. A stroboscope was used, together with a stone (a flat cylinder cut from a sheet of stone) with a black and white pattern painted on to permit measurement of the spin rate. This study also examined stone bounces on sand surfaces. As we discuss later, similar experiments are readily accomplished with a modern video or digital camera, with or without a strobe.

Bocquet (2003) applied some simple analytic expressions to explore the dynamics of a skipping stone, using a square flat plate and a circular flat plate as models. This work uses a simplified approach (e.g., assuming normal and side-force coefficients for forces generated by the water which are invariant with angle of attack) which compromises its accuracy, but it nonetheless captures the essence of the overall process. The work identifies the existence of a minimum speed V_c to skip, although this is restricted to a fixed (small) incidence angle. An approximate expression yielded by this analysis is

$$V_c \sim \left(\frac{15\,\mathrm{Mg}}{4\mathrm{C}\rho_1 a^2}\right)^{1/2},$$

with M the mass of the stone, radius a, g the acceleration due to gravity, and ρ_l the density of the liquid. C is a composite force coefficient (depending on angle of attack), but approximately the lift coefficient for small angles. Substituting some typical numbers $M \sim 0.1\,\mathrm{kg}$, $a \sim 0.05\,\mathrm{m}$, $C \sim 0.1$, $g = 9.8\,\mathrm{ms}^{-2}$ and $\rho_l = 1000\,\mathrm{kgm}^{-3}$, we find $V_c \sim 4.5\,\mathrm{ms}^{-1}$, a not altogether unreasonable result.

This work also discusses the need for spin, and suggests a limit on the number of bounces before the spin axis is unacceptably precessed. Due to the (assumed constant) energy loss at each collision, another limit N_c on bounce number can be derived, and the result follows that the horizontal distance $\Delta X(N)$ traveled during the Nth bounce varies as

$$\Delta X(N) = \Delta X(0)(1 - N/N_c)^{1/2}.$$

Further analysis required experimental support to determine which assumptions might be acceptable in the theory. Clanet et al. (2004) used high-speed photography with an aluminium disc as a "stone" to determine the effects of impact conditions on skip performance. A machine was used to launch the stone onto a water tank with specified parameters. The "stone" was spun with an electric motor which was itself mounted on a frame which could slide along a set of rails. The rails allowed the speed and flight path angle of the stone to be modified: the motor's orientation could be adjusted to modify the spin axis of the stone. A high-speed video system (~ 130 frames per second) was used to observe the stone's motion.

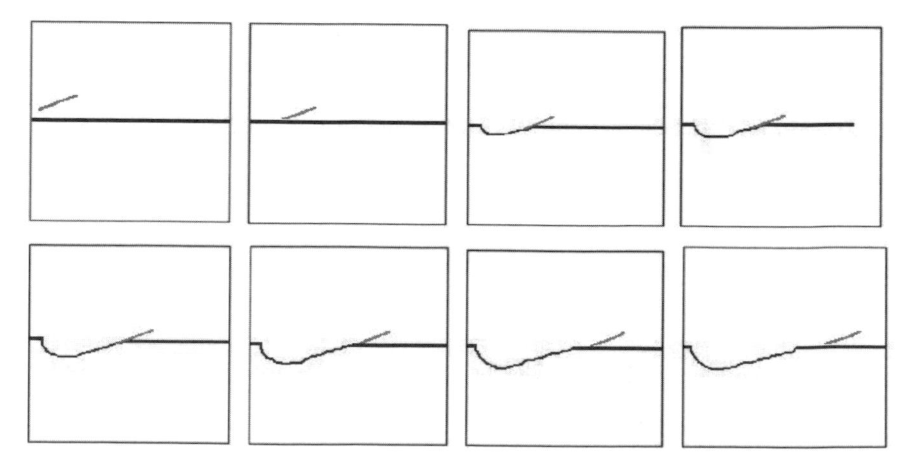

Figure 13.8. Interpretive cross-section of video record from Clanet et al. (2005). Snapshots 6.5 milliseconds apart, running top left to right then bottom left to right. The cavity formed in the water is highly asymmetric.

Although this work represents the most systematic study of stone-skipping to date, it is of note that the spin rate of the "stone" is much higher than is typical for a hand-thrown stone—some 65 rotations per second, and their apparatus was only able to achieve flight-path angles (β) of 15° or higher, whereas stones thrown outdoors are likely to have much more grazing incidence angles.

Clanet et al. found that the most crucial parameter is the attitude (i.e., the inclination of the stone's spin plane to the horizontal). When $\alpha = 20°$ the flight velocity needed to skip was minimized. For a flight-path angle (β) of 20°, a stone would skip for $\alpha = 20°$ at only 2.7 m/s, while it needed to be traveling at 3.5 m/s for $\alpha < 7°$ or $\alpha > 42°$.

Additionally, a domain of attitude flight path angle (β) space was established within which skips could occur. The range of allowable β is widest for $\alpha = 20°$. When $\beta = 42°$, α had to be almost exactly 20° for skip to occur.

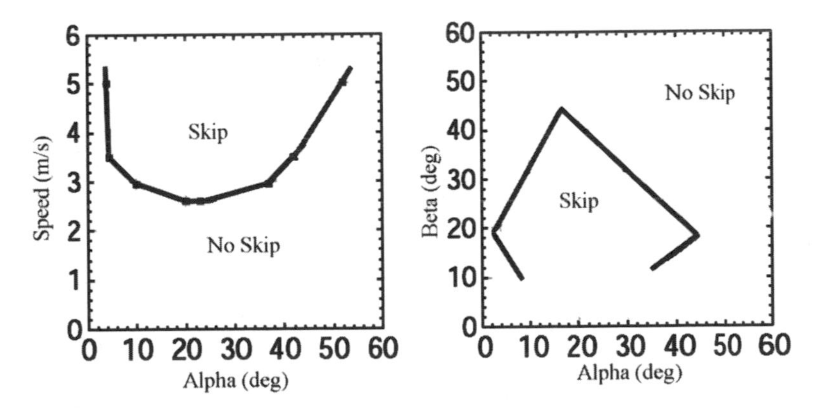

Figure 13.9. Stone-skipping envelope from the experiments of Clanet et al. (2004). These results correspond to an aluminium disc 5 cm in diameter, 2.75 mm thick, spinning at 65 revolutions per second. An attitude (α) of 20 degrees allows the widest envelope of speed and flight path angle (β) for skipping to occur.

One result noted by Clanet et al. (2004), but not explained, is the existence of a minimum contact time as a function of attitude β.

Some Simple Field Experiments

It is possible to learn much about the mechanics of stone-skipping using video or digital photography. These methods allow quantitative measurement of skip distance and some other effects. The images in this section were taken with a simple digital camera with an "action" mode that takes about 2 frames a second.

One effect that can be observed is the skip distance. In fact it emerges that Bocquet's simple model of skip distances following a geometric progression does not always hold. In fact, perhaps due to the uneven surface of the water, it can happen that one bounce is longer than the previous one.

Another important observation is that, for right-handed throws at least, a stone's later skips seem to veer to the right. Casual inspection shows that the distance between skips is typically ~ 80% of the previous skip.

Figure 13.10. Digital photo of a series of skips on Seil Sound, Scotland, with the ripple circles highlighted. The direction of travel swings to the right by about 30 degrees. Photo by Dr. Elizabeth Turtle.

Experiments with an Instrumented Skipping Stone

Exactly what happens during the few milliseconds of impact is difficult to tell from video records, which by providing position are effectively the double integral (i.e., a highly lowpass-filtered version) of the force history. To investigate the force history it is better to measure the force directly, or since a stone makes a rather good rigid body, its acceleration history. Thus some experiments that appealed to me were to install a "flight data recorder" on the stone—an accelerometer sensor and a data-logger to store its signal, just like the experiments with Frisbees and boomerangs.

A first attempt used a Basic Stamp II microcontroller storing two-axis accelerations at about 65 sample pairs per second from an Analog Devices ADXL210 accelerometer. This device measures +/−10g with a bandwidth of up to ~ 1 kHz, outputting the result as a 5 V pulse-width modulated signal. However the memory-write time of the microcontroller limits the sampling rate to much less. The microcontroller could also drive a flashing LED to permit streak photography to document the impact speed. The systems were driven by a pair of small NiMH batteries, and the data could be downloaded via a serial cable to a terminal program on a laptop computer. The equipment (weighing about 20 g) was attached to a flat stone by means of hot glue.

Despite the aesthetic appeal of the system, results were somewhat disappointing. The ~ 15 ms between samples was marginal for resolving the pulse shape, and the somewhat irregular mass distribution and aerodynamic drag area of the equipment on the stone led to considerable nutation during spinning flight. A major difficulty was the vulnerability of the equipment to interruption of function by water leading to electrical shorts, despite an effort to seal exposed conductors with glue.

Some results were nonetheless obtained, showing that the fundamental approach had promise. Results were interpretable, but faster sampling was needed to resolve the peak, a better mass distribution was needed for clean free-flight measurements, and a larger dynamic range of acceleration measurement was required.

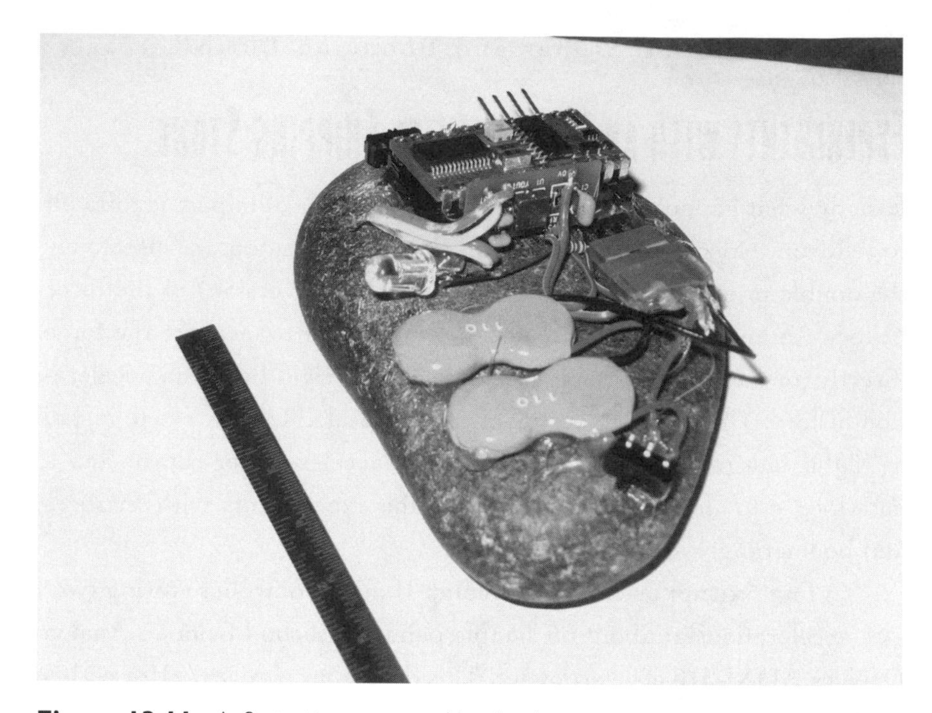

Figure 13.11. A first attempt at making in situ measurements of the accelerations on a skipping stone. This setup gave some basic data, but was dynamically poor, sampled the acceleration too slowly, and was not sufficiently robust or waterproof for prolonged experimentation.

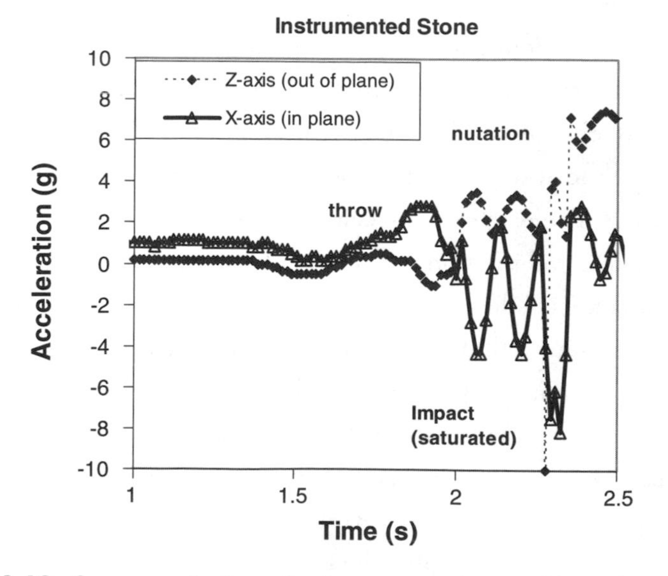

Figure 13.12. Some results from the instrumented stone. The spin modulation of the signals can be clearly seen, but the skip event is too intense and too short to be adequately resolved with this apparatus.

A second approach used a Crossbow CXL100HF3 accelerometer (+/−100g range in three axes, with a 0–5 V analog output). The three axes were recorded at 200 sample sets per second with 12-bit resolution by a Pace XR440 datalogger, a unit roughly the size of a pack of cigarettes and convenient for field operation (Lorenz, 2004). In this instance the data-logger was removed from its casing in order to fit into the package. A Speake FGM-1 single-axis fluxgate magnetometer provided a pulse-frequency signal proportional to the magnetic-field along the sensor axis: the output of this sensor was passed through a lowpass filter to generate an analog voltage which was recorded by the fourth channel of the data-logger. A separate 555 timer circuit strobed a set of high-brightness LEDs.

This equipment was installed in a small ($110 \times 110 \times 50\,\text{mm}^3$) "tupperware" container which could be closed with an airtight seal.

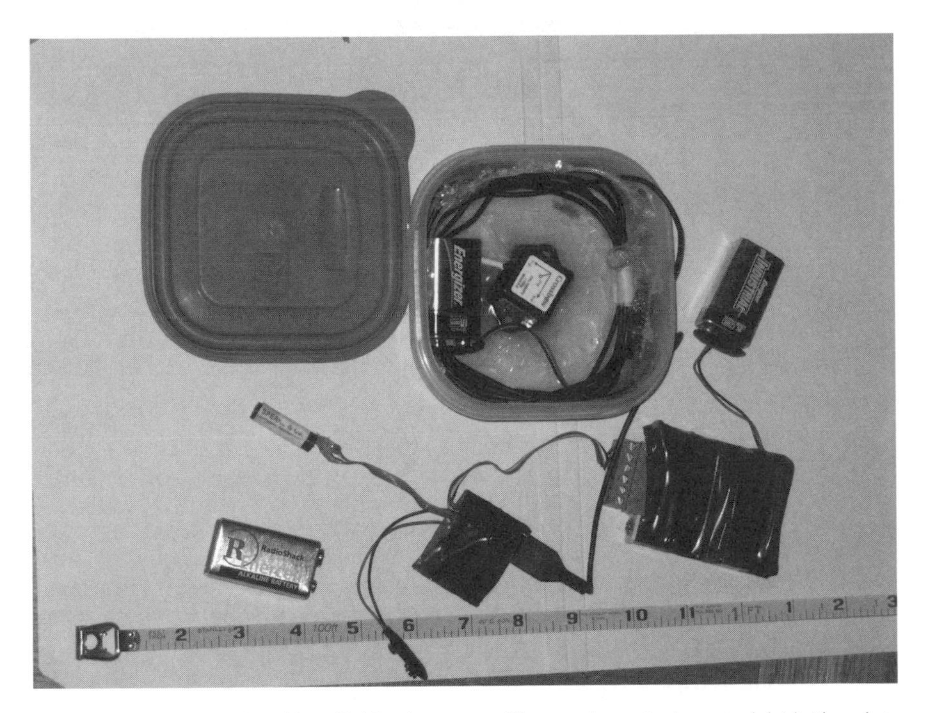

Figure 13.13. A "lunchbox" skipping stone. The package to lower right is the datalogger. The accelerometer is glued to the center of the box. At lower center and left are some power and signal conditioning electronics and the magnetometer. All of these parts can be fit into the watertight box.

It should be noted that this particular implementation was less dense than water, facilitating recovery. (It is straightforward to add lead shot to increase the density, which is an important parameter in the skip dynamics.)

The package was thrown in the usual way across the surface of a colleague's swimming pool (12 m long). A conventional video camera recorded the trajectory (adequate resolution to measure speed, although not attitude): the LEDs greatly facilitated tracking of the projectile in the video record. Positions were digitized using Videopoint software.

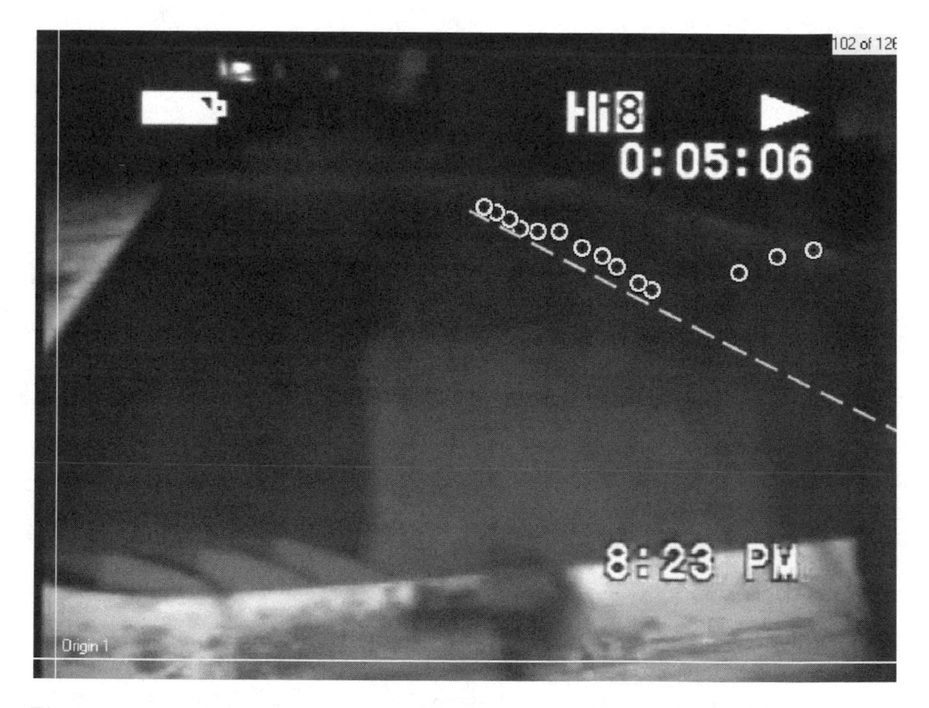

Figure 13.14. Conventional camcorder record of the lunchbox skipping stone. The positions were recorded with Videopoint software and are stacked to show the two approximately parabolic segments. The dashed line shows the "ground-track" of the stone—evidently the vertical speed after the first bounce is much less than before impact.

This video record (the yellow circles denote the stone position at approximately 40 ms intervals, with a few frames missing) shows clearly how the stone bounces much more shallowly from the surface than it impacts it (the dashed white line shows the "groundtrack" as a guide).

The package could record data at 200 Hz for around 30 s, enough time to seal the package with adhesive tape (to prevent the lid from bursting off) and throw. An example dataset is show below.

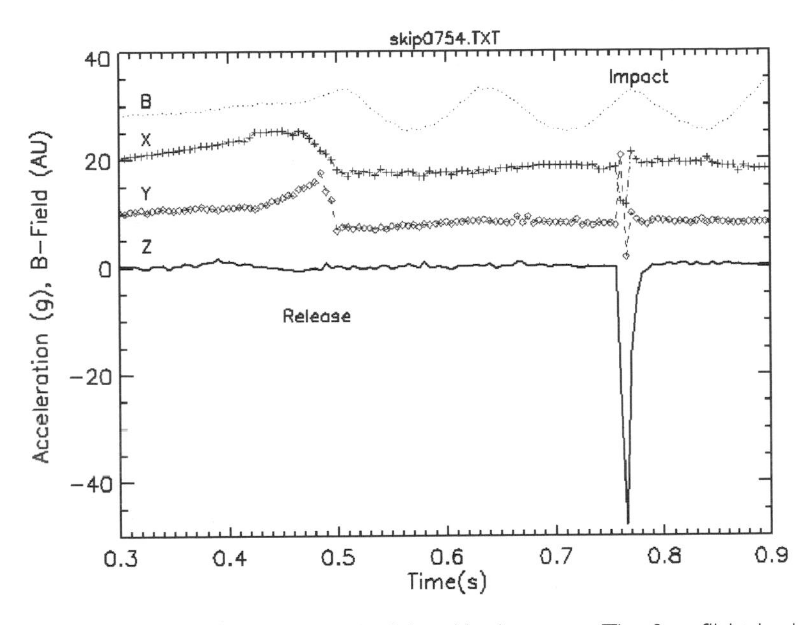

Figure 13.15. Acceleration record of the skipping stone. The free flight is about 1/4 of a second—during this time the stone makes two complete revolutions, as indicated by the magnetometer trace B. The *X*, *Y* acceleration traces are offset for clarity—they see more acceleration (as one would expect) during the launch, but their skip signature is 3 or more times smaller than the *Z*-axis acceleration at skip.

The acceleration signal is much cleaner, indicating spin around the axis of maximum moment of inertia, with the *Z*-axis coincident with the spin axis. Accelerations rising to ~ $10g$ for ~ 0.1 s are seen in the *X*- and *Y*-axes during the throw, indicating flight at ~ 15 m/s—this flight speed is consistent with the video record.

The magnetometer signal is spin modulated, as expected, indicating a spin rate of about 8 Hz (actually rather similar to a Frisbee—Lorenz, 2005). The spin slows slightly after the impact.

X and Y ("drag") accelerations are 10–20g. The Z-axis acceleration, which causes the bounce of the stone out of the water, peaks almost instantaneously at 48g, falling back to zero with an e-folding time of ~3 ms. This pulse shape is very characteristic of other liquid impacts.

Splashdown

In contrast to the simplified analysis of Bocquet (2003), the impulsive force on a body breaching the water surface is in fact rather complex. However, the resultant forces are reasonably straightforward to determine.

The analytical investigation of impact of bodies on the surface of water has a long history. The principal development was that of Theodore von Karman in 1929, in considering the impact of seaplane floats upon water. In essence, the problem is one of momentum conservation; as the float enters the water, it shares some of its momentum with it. Naturally, the momentum-sharing implies an equalization of the speed of the body of water with the impacting float. But how much is this "added mass" of water?

The problem received significant attention in the U.S. manned space program during the development of the Mercury and Apollo spacecraft, which were recovered by splashdown into the ocean. Scale model tests (e.g., McGehee et al., 1959; Stubbs, 1967) showed the accuracy of the added mass approach in computing the peak deceleration, which was usually the parameter of interest. More recently, there has been some work on splashdown dynamics by the present author (Lorenz, 1994; Lorenz, 2003), given the prospect of the landing of the European Space Agency's *Huygens* probe on Saturn's moon Titan in January 2005. That cold body may be partially covered with liquid hydrocarbons such as ethane, and a splashdown landing would be recognized by the deceleration history recorded by onboard accelerometers. (In the event, the probe landed on an area looking much like a beach or streambed; an area clearly geomorphologically modified by liquid, although the liquid itself was evidently elsewhere.)

Figure 13.16. Splashdown of the *Apollo 15* capsule. Notice the cavity formed in the water—the instantaneous diameter of the cavity is larger than the diameter of the capsule itself. NASA image S71-43543.

A not unreasonable estimate is of a hemisphere equal in diameter to the cross-section of the impacting object at the undisturbed waterline. More refined work simply scales this mass by an empirical factor. (The "added mass" concept is also used in the dynamics of airships and parachutes whose inertia appears to be larger than would be expected from the mass of the fabric itself—a certain mass of fluid is coupled to the fabric. You can experience added mass for yourself at the swimming pool: if you push yourself off from the side and allow yourself to drift and slow down, you will feel your wake catching up with you and pushing you along a little. The momentum in this wake derives in part from your effort in accelerating the added mass when you pushed off.)

When the impacting body has a simple algebraic depth-diameter relationship (a cone, for example, or a section of a sphere), analytic expressions (e.g., Hirano and Miura, 1970) for the added mass as a function of time, and thus the impact force as a function of time, can be developed. In the crudest terms, the increase of added mass with depth is a function of the radius of curvature of the body hitting the water—you can verify for yourself that the forces per unit area are much higher when diving into a swimming pool with a bellyflop (say radius of curvature of 1–2 m) than when diving head-first (radius of say 10 cm). It can be seen that the stone-skipping problem is largely the same. The only complication is that the entry velocity is not vertical, and the attitude is not horizontal. The added mass will (depending on how circular or square the stone is) vary something like $\sim \rho_l k S^n \partial^{3-n}$ where k is a constant and n some other constant $1 < n < 2$.

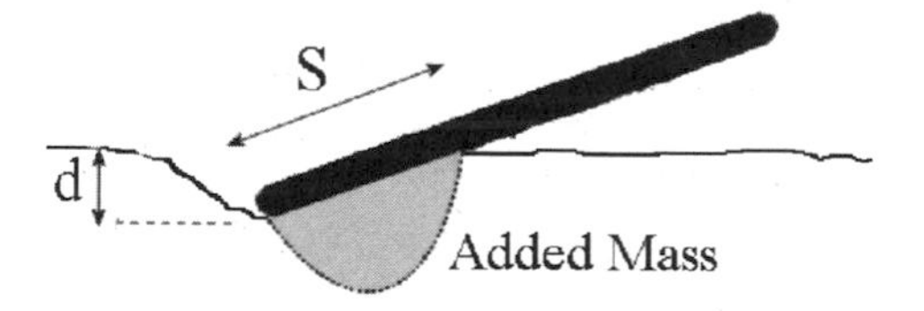

Figure 13.17. Schematic of the added mass (grey region) on a skipping stone.

Applying this theory to the instrumented skipping stone in Figures 13.13–13.15, we find quite reasonable agreement, with a peak of $\sim 50g$ in the vertical axis and $\sim 10g$ in a transverse one for $V \sim 12\,\mathrm{ms}^{-1}$, $\alpha \sim 10°$, $\beta \sim 8°$; $M \sim 0.3$ kg and diameter ~ 10 cm. Perhaps more crucial than the peak values, the fact that the acceleration rises near-instantaneously, with a subsequent decay timescale of ~ 3 ms, is very consistent with the "splashdown" model.

It is difficult to reconcile a formulation based simply on lift and drag coefficients and wetted area with this rapid rise and decay. This is not to say these forces do not play a significant part in the overall momentum budget of the event, only that they do not well represent the impulsive peak deceleration.

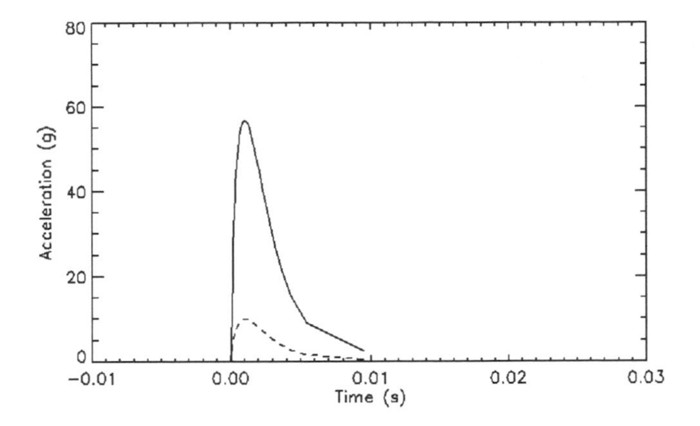

Figure 13.18. "Splashdown" model of the skipping event shown in Figure 13.15. The "added mass" approach yields a reasonable peak amplitude, the sudden rise time and fairly rapid decay of the *Z*-axis acceleration pulse (solid line) and the *X*-axis (dashed line).

One might question whether surface tension plays a role. In the direct sense, it does not—the reflection of a 200 g rock which spends, say 10% of its time in contact with the water, requires the exertion of about 20 N of force. This force has to act along a perimeter of the order of $\pi * 5\,\text{cm} = 0.15\,\text{m}$, or a tension of some 130 N/m. This dimension should be compared with the surface tension of water, namely 0.072 N/m. Thus surface tension can be neglected in the forces on the stone.

However, it can be readily seen from photographs that the water surface is substantially deformed by the impact, and builds to form a "ramp" in front of the stone. It may be that surface tension and viscosity play a part in controlling this aspect of the process, and other aspects such as the reduction in spin rate due to skin friction. It is clear that although key elements have been captured in the work of Bocquet and Clanet et al., and in the splashdown discussion above, much work remains before we fully understand the dynamics of stone-skipping.

Skipping Stone Aerodynamics

There is one situation that will be familiar to the experienced stone-skipper. A thin, flat stone which superficially appears ideal for skipping may, when thrown, roll anticlockwise in the air, cutting into the water surface rather than skipping on it. This behavior is of course simply understood as that of a "bad" Frisbee: the stone is acting as a wing as it flies through the air and develops an appreciable pitch moment. With an insufficiently heavy stone, spun insufficiently fast, this pitch moment precesses the spin vector over. The only solution is to throw with as much spin as possible, and at a low angle of attack (at zero angle of attack, the pitch moment is zero—however, this restricts the range of α with which the stone will hit the water).

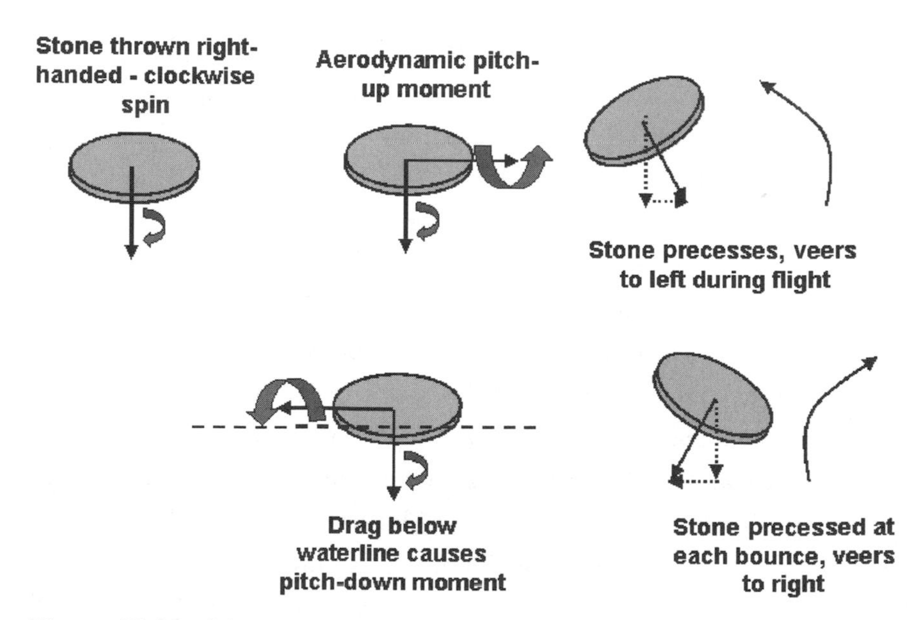

Figure 13.19. Schematic of the opposing effects of aerodynamic pitch-up in flight and the pitch-down on impact: the gyroscopic effect leads to veering left and right, respectively.

Suggestions for Good Skipping, and for Future Work

As with so many hand-launched objects, specifying the ideal throw and actually doing it are very different things.

Clearly, kinetic energy should be maximized, and this therefore means selecting a stone that is as massive as possible without slowing the throw (i.e., a mass comparable with that of the hand itself—much more than this and the throw will not be as fast). To minimize air drag losses, a dense stone is probably best.

The aspect ratio of the stone is a parameter which has not been optimized. Similarly, the spin rate may have an optimum—while obviously a large spin yields a high gyroscopic stiffness, it may be that the hydrodynamic forces and moments on a rapidly spinning stone introduce a pitch moment during the skip. A series of experiments with different aspect ratios (i.e., moments of inertia) and spin rates would elucidate this question. The optimum incidence angle appears to have been robustly determined by Clanet et al. as 20 degrees.

Interestingly, as I found with my own instrumented experiments, a perfectly circular object is not ideal—it is hard to develop enough torque during the throw to get the stone spinning suitably fast. A slightly oval stone may be better (of course, too narrow is bad, as the transverse moments of inertia become too dissimilar for stability) or even one with some corners. Jerry McGhee even makes stones of his favored shape (out of clay, such that they dissolve in the water to minimize any environmental impact!).

The attitude dynamics of a skipping stone, and specifically the precession of the spin axis on contact with the water, is an area that is yet to be fully investigated. One intriguing possibility is that a flight may exhibit the veering due both to the pitch-down moment on impact into the water and the pitch-up aerodynamic moment during flight between skips. Finally, the detailed behavior of the water surface during the skip, and its possible influence on the exit parameters of the stone from the skip, remains to be fully explored.

The Basilisk Lizard

There are a number of animal species that live on the water surface. At progressively smaller scales, the effects of surface tension become more and more significant. This is most evident in those insects that live on the surface of the water, such as pond skaters, water striders, etc. Indeed, a dimensionless number can be defined—informally referred to as "the Jesus Number," the ratio of surface tension force to weight ($Je = \gamma/\rho l^2 g$, where γ is the surface tension and l the characteristic length scale of the body, density ρ), which determines how easy it is to "walk on water." (Vogel, 1988).

For pond skaters etc., this quantity is large, and thus the insect's weight can be supported by the tension of the water surface. The weight of larger animals simply tears through the skin of the water. Remarkably, however, it is possible to beat this number (in a somewhat analogous way to the oft-quoted, but somewhat misleading assertion that bumblebees defy the laws of aerodynamics) by invoking nonsteady conditions. Here, the supporting force is nonsteady acceleration of water masses, rather than the static tension.

The best-known animal in this environment is the basilisk lizard, *Basilisus basilisus*, or sometimes less formally the "Jesus Christ lizard." Although unlike its namesake, it cannot *walk* on water, it can at least *run* across the surface of the water (e.g., Glasheen and McMahon, 1996a). It stops itself from sinking by transferring its weight (i.e., a continuous momentum flux) to the water by a rapid series of slapping steps. Its flat, splayed feet are slapped onto the surface during a rapid run, and in the process, net downward velocity is imparted to the water. The nonsteady impulse generation has only recently been elucidated by Hsieh and Lauder (2004) using particle image velocimetry with a laser light sheet to measure the motion of tracer particles in the water.

The lizard tends to lean forward while running, such that the force from its legs is expressed slightly in a forward direction, propelling the lizard forward across the surface of the water. As one might expect from size scaling of mass and area, small lizards are more easily able to run

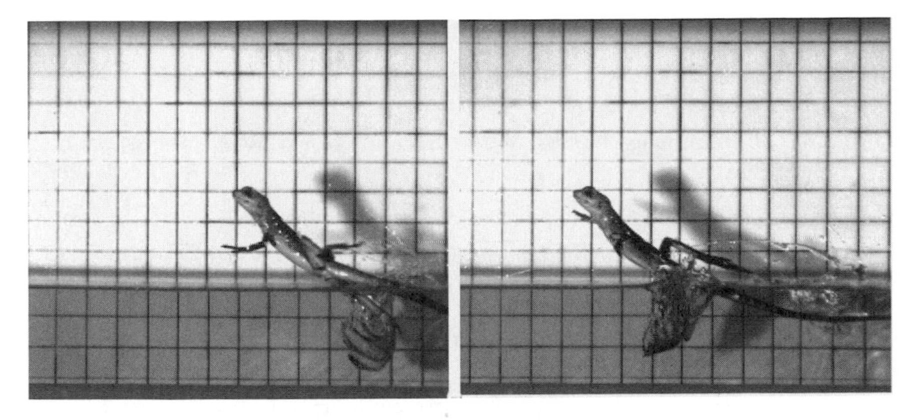

Figure 13.20. The basilisk lizard caught on the run in high-speed video. Note the arms flailing for balance, the cavity made in the water by the foot, and the location of the tail. Images courtesy of Tonia Hsieh of Harvard University.

on water and can generate substantial force surpluses, while fully grown (0.2 kg) lizards can barely support their weight (Glasheen and McMahon, 1996b).

In this instance, of course, the sustained attitude to permit multiple contacts with the water surface is maintained by neuromuscular control of the gait, rather than by spin-stabilization. Additionally, its long tail, which video shows remains immersed throughout each step, must be a major contributor to stability (as well as providing, perhaps, a little hydrodynamic lift).

References

Bocquet, L., The physics of stone skipping, *American Journal of Physics* 71(2), 150–155, 2003.

Clanet, C., F. Hersen, and L. Bocquet, Secrets of successful stone-skipping, *Nature* 427, 29, 2004.

Glasheen, J. W., and T. A. McMahon, A hydrodynamic model of locomotion in the basilisk lizard, *Nature* 380, 340–342, 1996a.

Glasheen, J. W., and T. A. McMahon, Size dependence of water-running ability in the basilisk lizard *Basilicus basilicus*, *Journal of Experimental Biology* 199, 2611–2618, 1996b.

Hirano, Y., and K. Miura, Water impact accelerations of axially symmetric bodies, *J. Spacecraft and Rockets* Vol. 7, pp. 762–764, 1970.

Hsieh, S. T., and G. V. Lauder, Running on water: Three-dimensional force generation by basilisk lizards, *Proceedings of the National Academy of Sciences* 101, 16784–16788, 2004.

Hutchings, I., Bouncing bombs of the Second World War, *New Scientist*, 2 March 1978.

Hutchings, I., The ricochet of spheres and cylinders from the surface of water, *International Journal of Mechanical Sciences* 18, 243–247, 1976.

Johnson, W., Ricochet of non-spinning projectile, mainly from water. Part 1: Some historical contributions, *International Journal of Impact Engineering* 21, 15–24, 1998.

Johnson, W., Ricochet of non-spinning projectile, mainly from water. Part 2: An outline of theory and warlike applications, *International Journal of Impact Engineering* 21, 25–34, 1998.

Lorenz, R. D., *Huygens* probe impact dynamics, *ESA Journal* vol. 18, pp. 93–117, 1994.

Lorenz, R. D., Splashdown and post-impact dynamics of the *Huygens* probe: Model studies, pp. 117–124, in *Proceedings of the International Workshop on Planetary Entry and Descent Trajectory Reconstruction and Science*, Lisbon, October 2003, ESA SP-544, European Space Agency, 2004.

McGehee, R., M. E. Hathaway, and V. L. Vaughan, Jr., Water-landing characteristics of a reentry capsule, NASA Memorandum 5-23-59L 1959.

Soliman, A. S., S. R. Reid, and W. Johnson, The effect of spherical projectile speed in ricochet off water and sand, *International Journal of Mechanical Science* 18, 279–284, 1976.

Stong, L., The amateur scientist, *Scientific American* 219, 112–118, 1968.

Stubbs, S. M., Dynamic model investigation of water pressures and accelerations encountered during landings of the Apollo spacecraft, NASA TN D-3980 1967.

Sweetman, J., *The Dambusters Raid*, Cassell, 2002. (First published by Janes in 1982 as *The Dams Raid: Epic or Myth*.)

Vogel, S., *Life's Devices*, Princeton University Press, 1998.

Von Karman, T., The impact of seaplane floats during landing, NACA TN-321, October 1929.

http://lpmcn.univ-lyon1.fr/~lbocquet/ricochet.html

www.stoneskipping.com

http://www.oeb.harvard.edu/faculty/holbrook/biomechanics/

14 Conclusions

It is not the purpose of this book to draw grand conclusions from its desultory survey of spinning flight. However, it is worth recapping a few general principles and comparisons.

A first general observation is that the dynamics of spinning flight has many areas that can benefit even from rudimentary further experimental investigation. The precision field of spacecraft attitude control, and the rough-and-ready but nonetheless exacting needs of weapon systems are well studied, but will continue to require new work. The dynamics of sports, however, has many unexplored areas. This book has attempted to provide references as a starting point for serious research, via dynamical simulation, computational fluid dynamics, or experiment, and even where publications exist on a particular topic, it is by no means certain that these are or should be the last words on the topic.

A second observation is that many aspects of spin are not well understood even by individuals working on them. The lesson of *Explorer 1* is a great example, and the spin of parachute-borne planetary probes seems to be something that is hard to get right.

A further point is that while many aerospace systems in their final forms work well and as planned, demonstrating during their development that they will do so can pose all sorts of challenges. The attitude motion of a spinning shell flying at several times the speed of sound is not a trivial thing to measure, nor is demonstrating the performance of yo-yo weights or airbags on Earth easy to do. Similarly, while the theory for measuring asteroid or comet spin dynamics from light curves is well established, it must be remembered that each data point being fit by the curve results from someone's hard work, often at a cold mountaintop observatory. Each data point is an observation that could have been lost due to bad weather or an equipment failure, and so the achievement of a useful result at the end should not be underestimated.

Finally, while spin is often taken rather for granted, closer examination shows that spin can be introduced for many different reasons: for stability against external torques, to suppress torques from misaligments and irregularities, to scan a sensor, to attain a desired curving or straight trajectory, or to slow a fall by autorotation. In some cases considerable effort must be expended to attain a spin, or to moderate it, while in others the desired spin is inevitable and predictable.

In terms of classification of flying shapes, there seems to be an almost unbroken continuum of objects from samaras to boomerangs to flying rings and discs. The same principles apply throughout—the pitch moment must either be exploited, as in the turning boomerang or samara, or suppressed or absorbed, as in flying rings and Frisbees.

Exploring these topics has led me to think a little more deeply every time I toss a Frisbee or skip a stone. I am not sure I can do any of these things more skillfully than before I started researching this book, but somehow they now seem, even with no more effort than before, more like rewarding experiments than idle pleasures. I hope you find the same.

Appendix 1
Instrumented Flight Vehicles

Although miniature flight instrumentation has been possible for decades, it is only in the last few years that it has become relatively straightforward and accessible to "amateurs." Two principal developments have helped. First are micromachined silicon accelerometers: their use in automotive airbag actuation, and subsequently in video game controllers has made them inexpensive, mass-produced items. Second is the easy-to-use microcontroller. Although microcontrollers have been around for some time, their general availability together with straightforward high-level programming tools (e.g., BASIC interpreters or compilers) is a more recent development.

Other miniature sensors (e.g., magnetometers) have also fallen in cost and size, and a variety of microcontrollers have become available, stimulated in part by interest in hobby robotics. Many vendors have sprung up offering microcontrollers, motors, sensors, and other systems that will be of interest if you plan to make flight experiments.

My first application of these devices was to record the dynamics of small parachute-borne packages (Dooley and Lorenz, 2004), in an attempt to gain familiarity with how accelerometer records could be used to reconstruct a gust environment. This is the problem encountered with parachute-borne planetary probes such as *Huygens*.

After these experiments I had the idea of trying one on a Frisbee, and thus the research that led to this book began. (I have since discovered that others had independently conducted similar experiments, although perhaps with different goals). Installation on a Frisbee introduced some particular challenges, in particular in reducing the mass (principally the batteries) and profile of the equipment. Additionally, the rapid spin required rather fast sample acquisition and storage.

I have since used similar flight-logging equipment on a balsa-wood airplane (the mass of the clean airplane is 20 g, but it flies acceptably with a 10 g instrument package). One possible future project is to use a small radio-controlled model aircraft to perform boundary layer meteorological measurements.

I've used two types of microcontrollers in these experiments: the Parallax Inc. Basic Stamp 2 (BS2) microcontroller and the NetMedia Inc. BasicX BX24. The differences are as follows:

- the BS2 is simpler and easier to program;
- the BS2 needs only a low operating current (low enough that it can be driven by two lithium button cells (CR2032).

On the other hand,

- the BX24 runs faster (although for miniature data-logging applications, the limiting factor is the write-time to the EEPROM, which is essentially the same for the BS2 and BX24);
- the BX24 has more advanced commands and capabilities;

- the BX24 has analog-to-digital converters on eight of its input pins, making it more flexible for reading sensors other than PWM accelerometers.

The two devices have (at present) the same cost—$50 each—and are pin-compatible with each other, so it is relatively straightforward to graduate from one to the other. Both vendors provide experimenter boards and documentation (Parallax Inc. in particular has a very large range of educational material on its website). There are also different variants of both microcontrollers, some faster variants, usually with higher power consumption, some with larger memory and more advanced interfaces. This aeronautics book is not intended as a primer for electronics or microcontroller programming; however, investigators familiar with basic electronics and with a rudimentary understanding of programming should be able to follow the material hereafter and subsequently develop their own ideas.

Electronic Circuit

The function of the circuit and program is to record the readings of four sensors—two accelerometers, a sun sensor, and a magnetometer—and read them out as numbers to a serial port for capture on a computer.

In the old days, accelerometers gave only an analog voltage. However, many modern devices are optimized for easy interfacing with microcontrollers and avoid the need for analog-to-digital conversion. Given a (typically) 5 V supply, they yield a square wave signal which is pulse-width modulated (PWM) by the acceleration in two axes. If it senses zero *g*, e.g., if that axis of the accelerometer (there are two) is horizontal, or the device is in free-fall, the duty cycle of the output is 50%. The duty cycle increases by 12.5% for each *g* of acceleration, up to 2 in each direction.

A convenient device for these experiments is the Analog ADXL202. This uses a micromachined silicon bending beam to sense acceleration, and can be tuned to optimize bandwidth against signal-to-noise (two 10 nF capacitors and a 120 k resistor on the board set the pulse width and sensing bandwidth to appropriate values, the PWM period being about 1 ms), and draws only about 0.5 mA. The accelerometer itself is a tiny surface-mount (SMT) device (about 5 mm × 5 mm × 2 mm) and costs around $15 per unit. A small circuit board can be made to handle the (rather fiddly) SMT component. An alternative approach is to buy the accelerometer already mounted on a small evaluation board (ADXL202EB, usually ~$50).

Analog devices manufactures other similar devices. The ADXL210 is the same as the 202, but has a +/–10*g* range, giving it the span to handle launch accelerations that would over-range the 202 during Frisbee launches, and the high in-flight accelerations for boomerangs. The ADXL250 is a slightly larger device with a 50*g* range, but gives its output as an analog voltage rather than a PWM signal. A variety of other manufacturers also make accelerometers in a variety of types and packages.

Note that most solid-state gyroscopes made with similar technologies can only handle rotation rates of 300 degrees per second, making them too slow for the typical spin rates encountered in the experiments in this book, although again modifications can be made to "hack" the scale factor implemented in the device, and new devices are coming along all the time.

Note that many further construction details, and an alternative accelerometer option, is given in an article in *Nuts and Volts* magazine (Lorenz, 2004).

The circuit must be assembled in a reasonably robust fashion, given its operating conditions. It can be built on a printed circuit board, stripboard, or even hard-wired in a bare bones fashion (since there are only a handful of connections to make). Complete with batteries and switch, a stripboard version weighed around 28 g: a bare bones version can be under 20 g. These can be compared with the 175 g weight of the Frisbee—less than 20%.

I added an external LED as a diagnostic (though note that the BX24 incorporates two internal LEDs). The program strobes it rapidly when it is taking data, and turns it on full-time afterwards when the code is reading out the data. A dark LED would be indicative of a problem.

One approach, if you have a Basic Stamp breadboard, is to download the program to it on that, and transfer the chip to the Frisbee setup. However, if you want to tweak the program, this can be tedious and offers many opportunities to bend pins on the Stamp. What I did was to make a separate cable to link a 9-pin serial connector to pins 1–4 on the Stamp via a small header. Because the serial handling for downloading the data from the unit after the flight is easier, the data output is on pin 5 (P0)—a two-wire header (or two pins of a 5-pin) connects pin 4 (ground) and pin 5 to a serial connector. This connector is easily attached after the flight.

BS2 Program Operation

Upon switch-on, the BS2 starts program operation. It first tests to see whether an external connector has bridged (shorted) pins 6 and 7 together (pin 7 is set high and pin 6 is pulled low by a 100k resistor to ground, unless pins 6 and 7 are linked). This is simply a convenient way of having the program output sensor readings to a serial port for inspection in real-time for sensor calibration.

The BS2 has a 2K EEPROM, which must contain the program as well as any data. The program itself is quite short, and leaves about 1600 bytes of EEPROM memory space to store the data—if all N sensors (here 4) are recorded at the same rate, then this means $1600/N$ sample sets.

The BS2 reads the accelerometer with the PULSIN command, which returns the length of a positive pulse in the PWM output stream, measured in units ("ticks") of 2 microseconds.

The sun sensor is a small photodiode. This charges up a capacitor, taking a time to do so that is roughly C/I, where I is the photocurrent passed by the diode. A 3mm square silicon photodiode will pass a milliamp or two in strong sunlight, and so will charge up a 100nF capacitor in about a tenth of a millisecond. The charging time can be measured with a BS2 command RCTIME, which returns the time measured in clock ticks that can be a couple of microseconds long (or shorter in faster microcontrollers). Thus in full sunlight, the time would be 50. In darkness, the charging time would be inconveniently long (RCTIME would time out), so a resistor is added in parallel to "short-circuit" the photodiode and force a maximum charging time that is not too long.

A magnetometer (Speake FGM-1) is shown in the circuit diagram: this sensor outputs a 5 V square wave with a period proportional to the applied field—typically the frequency is 70 kHz. This can be converted into a useful number by counting the number of pulses in a fixed window, using the COUNT command. The length of the count window is chosen to get an adequate number of cycles—say a couple of thousand.

Now, numbers like the PULSIN output are 16-bit (2-byte) integer words. To store each reading for the sensors would not only require two EEPROM write operations (which are slow) but would also require two bytes of EEPROM. Adequate precision (about 2%) for this application can be had with only 8 bits of data. The numbers are therefore scaled to an 8-bit range (0–255) by division and subtraction. The program reads the sensors and performs this conversion and stores only the least significant byte of the word.

With all these sensors, there are a variety of ways of squeezing a useful measurement range (+/–2*g*, or darkness to full sun, or along the Earth'vs magnetic field or against it) into the range 0–255. Changing the timing components in the sun sensor or the accelerometer will change the relevant pulse times. Similarly, the count window can be altered, or several PULSIN readings summed together. Then division and subtraction is used to scale to the right range—usually some trial and error is needed, most conveniently done in the "realtime" mode.

If pins 6 and 7 are bridged by a jumper (as shown) or a switch, the program simply outputs the sensor readings as decimal ASCII numbers separated by commas to a serial port, pauses to blink the LED slowly, and repeats ad infinitum. This real-time or "calibration" mode is useful for testing the equipment and identifying calibration readings. In the more usual case, the readings are stored in EEPROM and the microcontroller quickly blinks the LED and repeats until the desired number of loops has been made (ultimately limited by the amount of EEPROM memory space). You can of course also change the program to always run in real-time mode, or never, if that suits your purpose.

Once the program has finished sampling (the BS2 writes 1600 samples in about 12 seconds—long enough for anything except a record-breaking Frisbee flight!), it reads out the data to the serial port as two columns of numbers separated by a comma. You can get the sample rate by watching the LED to determine the exact time of the record, and dividing by the number of samples (1600/*N*, or here 400).

The data can be captured by connecting the serial output to a PC serial port and reading the output with a terminal program (Hyperterminal is installed on most Windows PCs)—the settings have to be 9600 baud, 8 data bits, 1 stop bit, no parity, no flow control. (These details will in any case have to be worked out to upload the program to the microcontroller in the first place.) The data can be captured to a text file for analysis using a spreadsheet or more advanced tools. I use code in Research Systems, Inc.'s Interactive Data Language to read in text files and make the plots shown in this book.

The sensor readings are reported as 8-bit integers: to convert back to real-world values, a scaling relationship of the form.

$$\text{Acceleration } (g) = \frac{\text{Reading} - \text{Reading } (0\,g)}{\text{Reading } (1\,g) - \text{Reading } (0\,g)}$$

should be used, where the $1g$ reading is for that axis of the accelerometer pointed downwards, and the $0g$ reading for the accelerometer pointed horizontally. Corresponding relationships can be used for the sun sensor and magnetometer readings. Note that the Earth's magnetic field is not, in most places, horizontal, but dips substantially towards the ground.

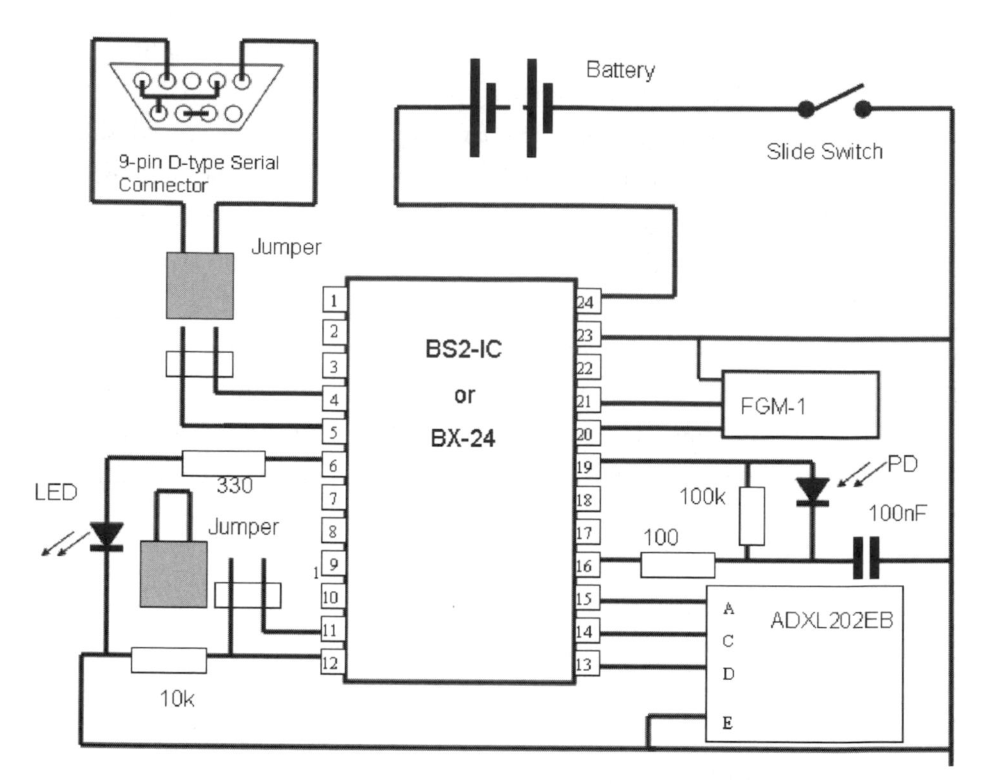

Figure A.1. Circuit diagram of a typical data-logger circuit (intended as a sketch of possibilities rather than a specific implementation). On the BS2, the serial cable shown is driven from pin 5 (P0) which requires a separate SEROUT command.

BS2 Program

```
' BASIC STAMP DATALOGGER
' FOR RECORD/TELEMETRY OF FLIGHT DYNAMICS DATA
' {$STAMP BS2}

t1x   var   word
t1y   var   word
sun   var   word
mag   var   word
i     var   word
```

```
start
' check run mode
' P7 high denotes real_time sampling and transmit
  for calibration
' P7 low denotes sample, store and subsequent
  transmission
' for convenience use P6 to supply 5V for test

high 6
input 7
if in7=1 then realtime
gosub Sample
goto Readout
end

Realtime:
' this loops forever, scanning the sensors and
  transmitting
' data is sent on P0 (pin5)
' LED on P1 (pin8) blinks on and off slowly

high 1
pause 50
serout 16,16468,[ 82, 32]
gosub Readsensors
low 1
pause 100
gosub Txdata
goto Realtime

Sample:
' samples accelerometer 800 times, saving low bytes
' for each channel in EEPROM
' sends stream of dots to serial line on P0 and
' rapidly flashes LED to denote status
```

```
for i=1 to 400
high 1
gosub Readsensors
low 1
serout 0,16468,["."]

write i*4, t1y.LOWBYTE
write i*4+1, t1x.LOWBYTE
write i*4+2, sun.LOWBYTE
write i*4+3, mag.LOWBYTE
next

' continues on to readout routine after sampling
  ends.

Readout:
' read out data
' sends row of asterisks to denote start of record
' then two columns of ascii numbers
' LED stays on
'

high 1
serout 0,16468,[cr]
for i=1 to 30
serout 0,16468,["*"]
next
serout 0,16468,[cr]

t1y=0
t1x=0
sun=0
mag=0
```

```
for i=1 to 400
read i*4, t1y.LOWBYTE
read i*4+1, t1x.LOWBYTE
read i*4+2, sun.LOWBYTE
read i*4+3, mag.LOWBYTE
gosub Txdata
next

' repeat in case data not acquired correctly
goto Readout

Txdata:
' write data as ascii to serial line in P0
serout 0,16468, [ dec t1x,",", dec t1y,",", dec
sun, ",",dec mag, cr]
return

Readsensors:
' reads pulse width from PWM accelerometer
' output. Does it twice for reliable results on
  ADXL202

pulsin 9,1,t1x
pulsin 9,1,t1x
pulsin 10,1,t1y
pulsin 10,1,t1y
' t1x and t1y are ~0.5ms (250 ticks) for ADXL202
' ADXL202 scaling relation - subtracting 80 seems
  to bring
' output into range 0-255
' t1x=t1x - 80
' t1y=t1y - 80

' read sun sensor
' first power up photodiode
```

```
high 14
' discharge capacitor
low 11
pause 1
' measure time and turn off
rctime, 11, 0, sun
low 14

' apply scaling to sun sensor
sun = sun \ 2 - 20

' read magnetometer
' by counting cycles on pin 15 for 2 milliseconds
count, 15,2, mag
' rescale to integer in useful range (adjust con-
stants ad hoc)
mag = mag \ 2 - 200

return
```

BX24 Code

The BX24 circuit functions the same, although the command syntax is different (the language is somewhat object-oriented and permits multitasking, neither of which feature I have exploited here). As mentioned before, the BX24 has analog-to-digital converters on 8 of its pins, which makes it easier to read infrared distance sensors, photodiodes, microphones, etc. It has some 32,000 bytes of EEPROM, permitting longer records e.g., for long boomerang flights. Another feature is that the BX24 has 400 bytes of RAM—this permits at least short records to be sampled at high rates (1–2 kHz) since the EEPROM write overhead can be avoided. I have, however, found the BX24 to be less robust in field applications, notably because the oscillator crystal can get ripped off the chip during collisions.

Note that different batteries have to be used, since CR2032 lithium button cells do not provide enough current for the BX24 which consumes about 25 mA. I have found small rechargeable NiMH batteries (Varta, 2.4 V 15 mAH) to work very well. Note also that the way the BX24 handles serial transmission is different.

For what it's worth, I have also used the BX24 in a data-acquisition application in a wind tunnel at NASA's Ames Research Center (Lorenz et al., 2005b) at pressures down to 20 mbar (to simulate early Martian atmospheres, only a little thicker than the one we see today at 6 mbar). The microcontroller seemed to be unaffected by the low-pressure conditions, although some ultrasonic range sensors worked less well at low pressure.

Other Microcontrollers

There are a wide variety of microcontrollers available. I have described the two that I myself have used, chosen largely for their ease of use (almost everything—voltage regulator, EEPROM, and basic interpreter—on the one 24-pin chip). There are many faster microcontrollers, cheaper microcontrollers (the PIC series is popular), and variants that may use programming languages such as C that electronics or computer enthusiasts may prefer.

References

Dooley, J. M., and R. D. Lorenz, A miniature parachute dynamics testbed, in *Proceedings of the International Workshop on Planetary Entry and Descent Trajectory Reconstruction and Science*, Lisbon, October 2003, ESA SP-544, European Space Agency, 267–274, 2004.

Lorenz, R. D., Frisbee Black Box, *Nuts and Volts* Vol. 25 No. 2, pp. 52–55, February 2004.

Lorenz, R. D., Flight and attitude dynamics of an instrumented Frisbee, *Measurement Science and Technology* 16, 738–748, 2005.

Lorenz, R. D., E. Kraal, E. Eddlemon, J. Cheney, and R. Greeley, Sea-surface wave growth under extraterrestrial atmospheres: Preliminary wind tunnel experiments with application to Mars and Titan, *Icarus* 175, 556–560, 2005b.

Appendix 2
Photography

One can study the dynamics of spinning (or nonspinning) objects with a variety of photographic techniques. These may be classified as video, streak photography, and stroboscopic photography.

Video

Modern video cameras are of course cheap, easy to obtain, and easy to use. The quality of conventional video can be radiometrically poor (i.e., the brightnesses are not well calibrated), but this is not usually a problem for kinetic measurements where only the position in the image plane is required. We have shown in this book several examples where simple video has been used (e.g., the Frisbee trajectories and the ball-in-cylinder motion). More sophisticated high-speed video can of course give better results, but can cost orders of magnitude more.

If using a video camera, don't forget to illuminate the scene adequately to keep the exposure time (and thus motion blur) to a minimum. Also, don't ignore the sound recording capability of a camcorder. Not only is a voice-over a good way of documenting an experiment, far easier than taking notes, but it stays securely attached to the relevant video file, which can be important if experiments stretch over a long period and notes can get mixed up. Also, there are ways of encoding additional information into the sound signal—even just the microphone can record a bounce, but one might build a circuit that can make a tone proportional to flowspeed or some other measured parameter.

A variety of software is available to measure positions in video frames or digital images (NIH Image, Image, Videopoint, etc., although software availability changes all the time—a web search may find other packages). It will be easier to translate the (x,y) pixel positions of image features into real-space coordinates if you make sure fiducial points like rulers or other markers are placed in the scene in which the moving object flies. There is much discussion of video techniques for kinematics in the biomechanics and physics education literature (e.g., Benenson and Bauer, 1993).

Strobe Photography

The technique of stroboscopic photography has always been somewhat miraculous in "freezing" rapid motions. The key is the short duration (a few microseconds) of the flash—this comes from a tube usually containing xenon gas which is broken down into a bright plasma by a high-voltage discharge. A circuit develops a high voltage to charge up a capacitor, which is discharged through the tube at regular intervals. Good stroboscopes with precision clocks operating over a wide range of flash rates are available for one or two hundred dollars. Alternatively, you could build your own from published plans (e.g., Sullivan, 2005) or from a kit.

Rather simple strobes and kits can be available for only $20 or so: these simply charge a capacitor at a rate determined by a variable resistor and discharge through the tube when a threshold voltage is reached. These are not precision instruments (though one could independently monitor the light level, say with an oscilloscope connected to a photodiode, to derive *a posteriori* timing information) but a more severe limitation may be that only strobe rates of up to 10–15 Hz are available. Nonetheless, this is enough for some applications.

Something that may be worth exploring, now that light-emitting diode illuminators are now becoming widely available, is to use these as a light source. A microcontroller or oscillator circuit could drive such an illuminator (without requiring high voltages, although a transistor buffer would need to be used to drive appropriate currents).

Streak Photography

Streak photography is the technique of attaching a light source to the object and holding open the camera shutter for the duration of the event concerned: the moving object forms a trail. This streak through the sky may have a particular shape that can be fit with a computer model, or may have geometric characteristics that are directly diagnostic of the dynamics (for example the cycloidal patterns in Figure 1.5).

However, a simple streak yields no time information. One way of adding this information is by shuttering. A method used to measure the speed of meteors is to place a spinning disc in front of the camera: the disc incorporates a slot, such that the exposure is interrupted once or twice per revolution.

Another approach is to have a modulated light source. The idea of installing strobing lights on a moving platform is of course not new. Felix Hess in his boomerang experiments used what he called a "time pill", a transistor multivibrator circuit, to flash a filament bulb with a period of 0.5 s. Driven by two 1.5 V batteries, this 3 g epoxy-encapsulated circuit delivered 195 mA to the lamp with a duty cycle of 80% (0.4 s on, 0.1 s off). This circuit allowed his photographs of boomerang flight to be given a time base.

Modern electronics makes such circuits rather easy to construct in three ways. First, many integrated circuits are available to construct oscillators with the addition of only a couple of discrete components (usually a resistor or two and a capacitor), rather than requiring the 16 components and all the associated solder joints needed by Hess's circuit. Second, modern ultra-bright light-emitting diodes (LEDs) can give adequate light output with lower currents than required by a filament bulb; they also have near-instantaneous response, permitting flash rates at hundreds of times a second or more. They are furthermore more robust than filament bulbs. Finally, the proliferation of cellphones and other mobile devices has pushed battery technology such that very small cells are available which can nonetheless supply high currents.

An oscillator is a very generic circuit and an enormous range of design possibilities exist. There is even an 8-pin device designed expressly as an LED flasher (the LM3909). Although a workable tracking flasher can be made with this device and a capacitor, it should be borne in mind that the device is really aimed at ultra-low power consumption (to flash an LED as a beacon, powered by a battery for months, for example to locate a light switch or flashlight in a dark room).

An additional bonus is that high-brightness LEDs are available in a variety of colors, making it possible to light different parts of the vehicle red, blue, green, etc. If a color image is used to perform tracking, then the trajectories of different parts of the vehicle can be isolated (e.g., the tip of the arm of a boomerang, and its center).

A circuit we have used with success is based on the popular 555 timer IC. This 8-pin device is very inexpensive, and with the addition of two resistors and a capacitor it can flash a set of LEDs with an arbitrary duty cycle and period. The IC is able to source or sink up to 200 mA, and so no driver is required.

The choice of current-limiting resistor is important. LEDs have limits on their drive current (specifically, this affects the junction temperature, and differential expansion of the encapsulating epoxy and the bond-wires can lead to failure). An LED can be driven at a high current, and will fail in a fraction of a second. Or it can be run at a modest current and last for hundreds of thousands of hours. Depending on how often you are prepared to replace a burned-out LED, you might choose to get as much brightness as possible. Because it is a temperature effect, it is possible to use very high drive currents at low duty cycles, as long as the pulse period is short enough (small compared with the thermal time constant of the LED).

Note that (since photon energy, which scales inversely with wavelength, relates to the semiconductor bandgap voltage) the forward voltage drop of the LED will depend on its color—red LEDs need a voltage of about 1.8 V, while green LEDs need about 2.5 V and blue close to 3 V. For a given drive current and supply voltage, the resistor needed will therefore depend on the LED color. Note also that a 555 IC can source or sink up to 200 mA, enough for around a dozen LEDs (depending on the desired current per LED). If larger loads are to be driven, a drive transistor is easily installed.

Consider driving a single red LED with a 4.5 V power supply driving the circuit. The voltage across the resistor will be roughly $4.5 - 1.8 = 2.7$ V. If we wish to limit the drive current to a sustainable 30 mA, the resistor should be 100 ohms. Halving the resistor will double the current.

For the boomerang flasher experiment, the component values used were $R_1 = 1\,k\Omega$, $R_2 = 22\,k\Omega$, $R_3 = 47\,\Omega$, $C = 2\,\mu F$. The short-pulse LEDs were green, and the long-pulse LEDs were red. The resistor values above give a duty cycle of about 5%.

In applications where we have used a microcontroller to acquire sensor data in flight, the microcontroller can be programmed to flash

LEDs, too (indeed, it is useful to do so, purely to indicate the status of the microcontroller operation, regardless of any utility in tracking). Note, however, that the output pins of most microcontrollers (such as a Basic Stamp, or the rather cheaper PIC series) can only source about 20–30 mA: bright output from an LED and/or driving multiple LEDs usually requires higher currents, and so some sort of buffer or driver circuit (typically just a FET or a transistor) is needed. One possibility that using a microcontroller affords is that a coded sequence of light pulses can be used (e.g., flashes of length 1, 2, 1, 1, 3 ms long) to facilitate correlation between different LED trails, or even to encode additional information.

To capture the event of interest it is of course essential that the camera shutter be open (even if modulated by a disk) for the duration of the event. Conventional (film) single-lens reflex (SLR) cameras usually have the capability to hold the shutter open indefinitely. Middle- and high-end digital cameras (but usually not the cheapest ones) have the provision to do manual exposures, typically up to 10–15 seconds. Beyond 15 seconds, the dark current on the CCD detector adds significant noise to the image, and this becomes a practical limit (detectors used for astronomical purposes with longer exposures are cooled to reduce this dark current).

Even if the light source on the moving object is bright enough to be imaged, it is obviously important that the scene is not so bright that the long exposure saturates the image. Thus boomerang experiments (like those of Felix Hess) must be done in darkness—in practical terms, since most free-flight experiments require tens of meters of free space, this means outdoors, at night.

Another possibility is to artificially darken the scene by using a filter, for example to pass only infrared light. Since only a small part of sunlight falls in this part of the spectrum, a longer exposure is less likely to saturate. Obviously, the beacon on the flying object must emit at a wavelength passed by the filter; fortunately infrared LEDs are inexpensive, powerful and easy to obtain, being used widely in remote controls for consumer electronic devices.

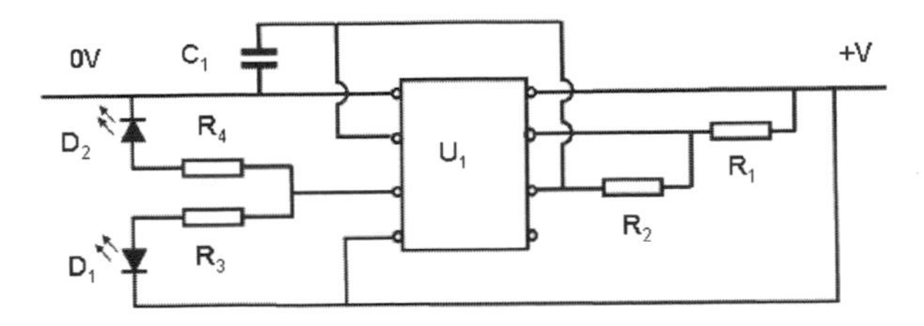

Figure A.2. Oscillator circuit for strobing LEDs—used in Figure 11.17.

V	3 to 9 V typical
U_1	555 timer IC
R_1	1 kOhm
R_2	20 kOhm $R_1/(R_1 + R_2)$ determines duty cycle
C	2 μF capacitor $C(R_1 + R_2)$ determines period
D_1	short-streak (10%) LED
D_2	long-streak (90%) LED
R_3, R_4	10–200 Ohm, depending on supply voltage, LED voltage drop, brightness/lifetime desired.

Reference

Benenson, W., and Bauer, W., Frame grabbing techniques in undergraduate physics education, *American Journal of Physics* 61, 848–852, 1993.

Index

Index

Index